Practical Systems Thinking

SYSTEMS WAYS
SYSTEMS PRINCIPLES

Kamran Eftekhari Shahroudi • Steven Conrad •
Jill Speece • Kirk Reinholtz • Martin “Trae” Span •
Sarwat Chappell • Quentin Saulter •
Golam M Bokhtier

Practical Systems Thinking

Finding leverage on complex problems

Springer

Kamran Eftekhari Shahroudi
Systems Engineering
Colorado State University
Fort Collins, CO, USA

Steven Conrad
Systems Engineering
Colorado State University
Fort Collins, CO, USA

Jill Speece
Industrial and Manufacturing Engineering
California Polytechnic State University
San Luis Obispo, CA, USA

Kirk Reinholtz
Systems Engineering
Colorado State University
Fort Collins, CO, USA

Martin "Trae" Span
Systems Engineering
Colorado State University
Fort Collins, CO, USA

Sarwat Chappell
Systems Engineering
Colorado State University
Fort Collins, CO, USA

Quentin Saulter
Systems Engineering
Colorado State University
Fort Collins, CO, USA

Golam M Bokhtier
Systems Engineering
Colorado State University
Fort Collins, CO, USA

ISBN 978-3-031-89393-3 ISBN 978-3-031-89394-0 (eBook)
https://doi.org/10.1007/978-3-031-89394-0

This Springer imprint is published by the registered company Springer Nature Switzerland AG
The registered company address is: Gewerbestrasse 11, 6330 Cham, Switzerland

We cannot solve problems by using the same kind of thinking we used to create them—Albert Einstein

Foreword

Discovering Systems Thinking Through Construction, Litigation, Hostage/Crisis Negotiation, Leadership, and Teaching by Zane Scott

"I wish I had known then what I know now." That's a refrain that has fallen more and more from my lips as I grow older. But nowhere is it more true than in the area of systems and systems thinking. My career has been a crazy quilt of seemingly unrelated learning curves leading to responsibilities in a variety of settings. From the construction work of my college summers, to the litigation of my days as a young lawyer, to the interpersonal skills of the hostage/crisis negotiator and mediator and into the management and teaching functions of the last 20 years, I have stitched this quilt together from the disparate pieces gathered in law school, court houses, crime scenes, and mediation rooms, classrooms, and conference centers. Only now, as I look back at my journey, have I begun to see the design of the quilt that I have labored to construct. In the last phase of my career, I had the good fortune to join a company led by three of the thought leaders of the systems engineering world, the President and founder of the company, David Long, his father and pioneering systems engineer, Jim Long, and the brilliant "lead dog" in my consulting engineering team, Joe Maley. David and Jim are the only father and son Fellows of the International Council on Systems Engineering. Each of them in their own very different ways taught me the foundational principles of systems thinking and systems engineering "done right." Together they had a profound effect on this extrovert who had chosen people over things at every opportunity. The fundamental ideas that tell us how systems are constructed and how they function became clear as they introduced me to the work. One of my first assignments was to breathe life into their idea of a "Primer for Model-Based Systems Engineering." Taking their knowledge, I wrote and David edited while all three of them answered questions and provided instruction. The result was a Primer that is still in use across the profession. But the big result for me was the birth of the idea that "systems" and "systems thinking" were valuable not just for making things—satellites, missiles, boats, and such. My lack of an engineering background freed me to see broader applications. The principles of systems thinking could be of great assistance in understanding and constructing social systems, organizations, laws, and regulations.

This insight arrived for me through their teaching and coaching which began 20 years into my 40-year career. As the light of systems thinking began to shine, I began to wish I had seen those concepts and principles earlier in my educational and career journey. I saw that, as a regulatory attorney enforcing the environmental and safety laws in Virginia's coal industry, I could have seen the systems that extracted the resources as part of the natural and community systems had I only understood systems thinking and design. In fact, I saw systems as playing a role in every piece of my "crazy quilt" career. If I had only known then . . . As the authors of this book point out, education in systems thinking should begin early—before career goals are set. That way systems thinking does not become tied to a narrow application. Systems thinking is a useful citizenship skill even outside the career space. For example, the city closest to my home is grappling with how to best serve a homeless population. The obvious, linear solution is to offer services—shelter, food, counseling, etc. But some people in the community oppose that approach, arguing that the better the services provided by the city, the greater the attraction for homeless people in the surrounding areas to move into the city adding to the burden of homelessness on their community. The truth contained in their objection could be understood by system thinkers to be a system boundary issue. Homelessness does not end at the city limits. The system with the problem is bigger than the city. The solution must be bigger as well.

A real "system solution" must address the greater system or the problem may simply move around the system. Any consideration of possible solutions must be preceded by an understanding of the system boundaries. That is fundamental systems thinking. Understanding systems thinking is part and parcel of our civic duty if we are to seek real solutions. When they craft solutions to the problems of homelessness, the city's voters need to understand the systemic nature of the problem and solution spaces. As this book points out, the concepts they need should be introduced at all educational levels and in the context of all disciplines. People who pursue any of the variety and levels of educational opportunities would then be exposed to systems thinking across a variety of problems. That way community problem-solving can benefit from systems thinking arising from a shared understanding of the fundamentals of the nature and behavior of systems.

The world of systems expressed in problems and solutions is rapidly evolving as technology and complexity accelerate around us. We need a grasp of the fundamental principles and tools for systems thinking to help us deal responsibly with their challenges both inside and outside the workplace. That means that systems thinking education must begin early in the educational journey and begin now. This book is a step forward to that beginning. The systems world would do well to read and heed its ideas, experiences, and proposals.

It can be tempting for experts to wait for the perfection of system science before teaching the evolving tools and principles. But even the tools and principles as we now have them are useful. If I had had them earlier, I would have been a better litigator, counselor, negotiator, mediator, leader, and teacher. I am thankful for what I have learned and those who have taught me along the way. But still, if I had known then what I know now . . .

Gate City, VA, USA
October 2024

Zane Scott

Discovering Systems Thinking Through Teaching, Research, Entrepreneurship, and Elected Office by Wade Troxell

When I was growing up, one of my favorite shows was The Lone Ranger. The show's main character and namesake was a wholesome hero who fought outlaws in the American West. At the end of each episode, when the Lone Ranger had dispelled the episode's central problem—thwarted the bad guy, saved a town from calamity, etc.—he would leave behind a silver bullet, which was his signature mark. In the show, the silver bullet was a representation of the character's purity, a symbol of honest and straightforward justice in a dangerous world. But the silver bullet had another connotation as well. It was a subtle wink and nod to the audience, serving as a reminder that the Lone Ranger was just a television character and that the problems he solved were, ultimately, simple. The phrase "there is no silver bullet" implies that a simple, straightforward solution does not exist to a difficult, real-world problem. In today's complex world, there are lots of hard, complex problems to solve and very few, if any, silver bullets.

Nevertheless, we frequently shy away from acknowledging complexity, both in real-world challenges and their potential solutions. This is readily apparent in our educational system, where we are taught to find well-defined answers that solve well-defined problems. In arithmetic, we are given two operands and an operator, say, $2 + 3$, to yield the answer 5. In geography, we memorize constitutive relationships, such as the capital of Colorado is Denver. These are well-formed problems with definite, describable answers that can be graded as being correct or incorrect. As a result, students often leave their formal education prepared for a structured, contained, predictable world, defined by known problems and clear paths to simple solutions.

In the real world, however, an answer to a given problem is not as apparent. This is especially true when complicating factors of the real world are present. Complicating factors can be people and individuals, including ourselves, as we are trying to solve such problems. Variabilities and their dynamic interactions, vague entities and transient relationships, uncertainty regarding appropriate scale and scope—ambiguity lingers in every corner of real problems. It is human nature to jump to answers, to shy away from ambiguity, and to oversimplify the world to reduce uncertainty. We tend to prematurely jump to what we think are answers before we even understand the problem. But, to truly understand the real problem, the problem as it exists, we need to confront it in its context and its circumstances—that is, in full view of its complexity.

What do you do when there is no silver bullet? How do we find non-simple solutions to non-simple problems? How do we even start? All is not lost. No need to be discouraged. Much like when we climb to a higher vantage point when we're lost, sometimes we need to take on a different perspective to see something new. Sometimes our mental models—our preconceived assumptions about how the world works—limit us. Sometimes we need to zoom out to see the bigger picture. The bigger picture helps us see other worlds and their relationships to a problem of interest. A systems point of view can help us see a problem differently, more holistically, even though it is right in front of us.

I have been a university professor for many years. I have worked in the private sector and served on boards of for-profit and nonprofit organizations. I have been an entrepreneur starting a technology company and have led a technology investment company. I have served as a city council member and mayor of a mid-sized US city. And in each, I have been responsible for charting a path through complex, ambiguous, unprecedented problems. Throughout my career, systems thinking has been a crucial skill I have used to navigate and address challenges in these wildly different sector roles.

As a focal point, in my role as Mayor, I adopted economic development and innovation as the unifying notion in what is described by Henry Etzkowitz as the Triple Helix. The Triple Helix is used as a metaphor that allows us to understand the non-linear interaction between the university, companies, and government. The Triple Helix integrates the sectors that generate knowledge (universities), those that use or consume knowledge (companies), and those that regulate and promote economic activity (government). These differing sectors are all aimed at generating and communicating scientific and technological knowledge. In concert together as a system, the Triple Helix model fosters the private sector to innovate and take to market solutions that in turn create economic development and increase the need for an educated workforce. In short, the system works as a positive feedback loop for the benefit of society.

By adopting a systems thinking approach to a complex world we can clarify, understand, and communicate what may otherwise appear dimly lit and too foggy to discern. A natural tendency is to zoom in and examine the individual pieces and parts, the constitutive elements, on the belief that understanding each individually amounts to an understanding of them all collectively. But a systems thinking approach reminds us that the whole is greater than the sum of its parts: to examine the system, we need to zoom out and look at the elements as to how they relate to each other, considering different perspectives for a more holistic viewpoint. When we zoom out, complexities can be more easily seen as interactions and interrelationships between components making up the system. In other words, sometimes we need to zoom out to understand the real problem before zooming in to solve it.

Whether you are a student or a professional, the first step to making an effective impact on our complex world is systems thinking. There is no silver bullet; there are no easy answers. The purpose of this book is to help you zoom out to zoom in on effective solutions. You will learn ways of understanding complex, real-world problems that lend insight into successful approaches and solutions. The contents of this book will help you command the use of systems thinking to confidently approach and better understand our dynamic, complex world.

Fort Collins, CO, USA
October 2024

Wade Troxell

Discovering Systems Thinking Through Film Production and Program Management by Andrew Williamson

I've produced approximately 50 productions during my career as a film and television producer, each a unique challenge requiring the coordination of countless moving parts. Every project was a complex system—scripts, actors, crews, locations, and unpredictable variables—all interwoven and interdependent. While I didn't have the terminology at the time, I now recognize that I was practicing systems thinking throughout my career.

In the fast-paced world of film production, seeing the big picture isn't just advantageous—it's essential. My role demanded an understanding of how changes in one area could ripple across the entire project. Adjusting a shooting schedule, for example, could impact the availability of actors, the logistics of equipment rentals, and even the morale of the crew. Success required both a strategic overview and a capacity to respond to details in real-time.

Looking back, I realize that if I had received formal training in systems thinking at the outset of my career, I could have navigated these complexities more efficiently. Instead of learning solely through trial and error—often under tight deadlines and high stakes—I could have applied established principles to foresee challenges and devise strategic solutions proactively. An understanding of this framework would have enhanced my ability to manage projects effectively from the beginning, increasing my odds of early success.

Systems thinking has come up often in my current role as a program manager for a national sustainability organization. Initially, I was put off by this concept as it was unfamiliar and I didn't understand how it applied to the work I was doing. I've come to realize that the skills I relied on as a producer align closely with systems thinking principles. Recognizing this connection has provided a fresh perspective on how I've approached complexity and problem-solving throughout my career, both in film and now in my work on sustainability and workforce development.

Burnaby, BC, USA
November 2024

Andrew Williamson

Preface

Despite excellent existing material in systems thinking theory and practice literature today, we still do not yet have an easier more practical path for the less experienced younger talent. Our goal here was to simplify and condense the systems thinking approach and insights embedded inside many references and case studies. The result is a small set of concise, practice-derived systems principles that are always true and improve our odds of success when we apply them to our complex real-world challenges. The central idea is that when the young talent reaches graduate school or enters a profession, they will no longer have to unlearn the linear or compartmentalized inside-the-box thinking fundamentals that hamper most of the workforce today. We hope that professionals who have already achieved an expert-level mastery of systems thinking philosophy, principles, and rigorous methods will find this material useful as they try to coach and push their organizations toward a culture that fosters systems thinking. This textbook benefits from the experience of teaching systems thinking and the rigorous tools that enable it for 15 years in addition to corrective feedback from 400+ graduate students ranging from entry-level to 25+ years of professional experience who applied it to real-world problems of their choice. Particularly, industry professionals taking the graduate class *SYSE 505—Systems Thinking for the Real World* taught at Colorado State University felt very strongly that we should teach this material to secondary and undergraduate students of *ALL* disciplines without a particular *technical* or *disciplinary* bias. We believe that this book provides a smoother and easier transition to learning and applying systems thinking than having to unlearn years of linear, inside-the-box disciplinary ways of thinking fostered by dominant organizational cultures years later. We wrote this book because we recognized the practical shortcomings of classical problem-solving, have a passion for systems thinking, and feel the urgency to boost the future generations.

Real-World Problems Are Becoming Increasingly Complex

Many people would agree that the real-world problems we face are becoming harder to understand, predict, or deal with: loss of life and capital due to forest fires, exploding healthcare costs, heavy traffic in every city (and in between the cities), pollution, homelessness, and cost and schedule overruns at best-in-class businesses or government organizations are just a few examples. These problems are worsening despite major efforts and expenditures specifically designed and aimed at improving them.

There are many reasons why systems operating in the real world are becoming more complex. One reason is that the needs and the available technologies to meet them are evolving. This causes the architectures of our systems and their interaction with their environment and other enabling systems to continue to evolve. This evolution in-turn causes emergent dynamic behavior that is increasingly harder to predict. There is also the human impact. Humans underpin the culture that drives the stated and true purpose of our systems. While automation reduces the direct involvement of humans in performing specific functions, such as steering a car, the number and severity of issues arising from the interaction of systems with humans and other man-made systems increases. For example, fully automated steering of automobiles causes many new and previously unknown safety concerns that humans and other systems operating in the same environment must contend with.

Large Number of Excellent Books and Articles on Systems Thinking

There are many insightful textbooks on systems thinking and the rigorous tools intended to enforce and implement it. There are many more articles and videos promoting Systems Thinking as the best way to view, approach, and understand complex problems. These references mostly say the right things and encourage better behaviors and thinking to approach real-world complex problems. Yet unfortunately, their key concepts and principles are very piecemeal, implicit, domain-specific, and distributed among a large volume or body of knowledge. Furthermore, the teachings are generally not targeted at young fresh talent from ALL disciplines. Many texts assume a background in or a willingness to invest in a disciplinary expertise like Systems Engineering to learn systems thinking. While others teach principles that are either not always applicable or, if misunderstood, can drive precisely the wrong kind of thinking and actions that make it even harder to understand real-world problems. So while we have learned from and respect the large body of knowledge on Systems Thinking, we believe this book is a necessary improvement to how we teach the *Systems Principles* and their *Application* to real-world problems.

The essential difference of this book is that it bases the teaching of systems thinking philosophy on a set of concise *Systems Principles* that are generically applicable to understanding real-world problems and the systems that operate in or interfere with them in some way. We intend to make this text unbiased toward any particular STEM or non-STEM reader background. Systems thinking is not just about terminology, buzzwords, or philosophy. It is also about rigorous methods that enforce and apply them. That is why we made it easier to get exposed to rigorous methods using real-world problem examples.

We need to clarify that we intentionally steered the material of this book more toward a *Systemic Understanding* of real-world problems and thinking of real-world problems as systems. We fully appreciate that one of the primary reasons for wanting to understand a problem is to create a solution for it! In other words, creating *Systemic Solutions* is the other side of the same coin that also has *Systemic Understanding*. However, we have not yet found a great way to remove disciplinary, domain-specific, and specialist knowledge bias when we talk about creating solutions for real-world problems. In this book, we give examples where better understanding leads to better solutions. Yet finding a way to teach creating *Systemic Solutions* without disciplinary bias to young fresh talent of all backgrounds is a hurdle that we have not yet managed to overcome. We hope that soon we or other systems thinking colleagues will create an easy text that will target creating *Systemic Solutions* in a way that is aligned with creating *Systemic Understanding* that we teach in this book.

Adoption of Systems Thinking Philosophy and Rigorous Tools

So why is the adoption of systems thinking (as measured by the rigorous tools that are intended to enforce it) so low despite the increasing number of complex challenges they face and repeatable fiascoes and blunders that impact even the best-in-class organizations employing experienced and reputable professionals? We believe the explanation lies in the interplay of the following factors:

- *Systems Principles*: Lack of a complete and concise set of principles that are easy to understand and apply to all disciplines and all complex challenges (despite the plethora of systems thinking textbooks, Handbooks, Articles, and Videos that attempted to fix this)
- *Classical Education*: Lack of systems thinking as a major element in secondary and undergrad education (versus top heavy graduates and working professionals)
- *Language and Culture*: Lack of systems thinking in our language and a culture that fosters and rewards linear compartmentalized thinking
- *Systems Theory*: Lack of theory or laws that explicitly underpin the empirically observed *Systems Principles*

The word "lack" here is not meant to offend the many excellent individuals and organizations, many of whom we know professionally, working tirelessly on improving the above situation. Indeed, if a keen learner invests enough time and resources in reading all the great references, videos, and movies on systems thinking, collects the "gems," takes classes on systems thinking philosophy, learns rigorous tools like system dynamics, implements them to solve their daily and long-term challenges at work or in their study, reflects and integrates all the insights from the above activities, then yes they will feel very empowered. They will have a significant record of success as they approach complex problems.

This is the arduous path that the majority (if not all) systems thinking experts today have followed. Unfortunately, this is the slowest and least practical way for mass adoption of systems thinking because:

- *Resources*: Impractical investment of time, effort, and money are required to come up the learning curve.
- *Inertia*: Auto-immune reactions and inertia of dominant linear thinking organizational culture create difficult hurdles.
- *Inflexible Mindsets*: The learning process inevitably involves reversing or unlearning fundamentals that were deeply ingrained in secondary, undergraduate education and years of professional experience.

A typical result of the arduous path to systems thinking has been to allocate the responsibility for systems thinking to a small subset of roles (e.g., systems engineers), instead of realizing that all critical functions (engineering, management, marketing, sales, finance, contracts, etc.) have significant systems thinking roles to play.

Mass adoption occurs when systems thinking becomes a significant portion of secondary and undergrad education in all fields where learners are exposed to "real-world" challenges that do not fit perfectly inside one of the main STEM or social science disciplines. There is no animosity toward classical disciplines. We need ways of better leveraging them together as we face real-world complex problems so that we can identify the unwanted side effects or adverse emergent behaviors when our solutions are implemented in the real world.

The main goal of this text is to provide an easier path for the young and fresh talent of all disciplines to adopt a systems thinking mindset, ideally before linear compartmentalized thinking dominates their mental models. We hope this will help the severe shortage of systems thinking talent in many US industries particularly those located away from the West and East Coasts.

Next, we share some of our (i.e., the authors') personal real-world observations and experiences related to systems thinking. It explains why we adopted systems thinking and what motivated us to teach it to others.

From Industry Challenges to Creating a Systems Thinking Talent Flow Pipeline

Kamran Eftekhari Shahroudi, PhD, INCOSE ESEP

Systems Fellow (Woodward, Inc.) and Professor of Systems Engineering (Colorado State University)

As a species, we have always faced complex problems, but more recently, our ability to create new problems appears to be outpacing our ability to understand, predict, and react to them. Climate change, runaway growth in US Debt, the adverse psychological impact of social media on children, and the adverse social impact of globalization are but a few among many real-world examples. There are many counter-intuitive examples where increased effort and expenditure have made the problems worse, not better, like homelessness, health, education, and innovation rankings in the USA. Often, the scale of the problem is such that we cannot easily spend or brute-force our way out of it. We need ways to understand real-world problems better, define a specific problem to go after, and identify the best points of leverage on the specific problem.
Systems thinking is the best way to understand and deal with complex real-world problems. However, the ability to view and think about real-world problems as systems is not something that most of our fresh talent naturally develops. We need an easier path to get started with systems thinking, get training on the rigorous methods that enforce it, and create a talent flow pipeline of systems thinkers into the workforce. The next three subsections share my experience on these topics.

Getting Started with Systems Thinking

After only a few years at Woodward, a 150-years-old American company focused on energy control solutions of land, marine, and aerospace systems, I noticed that despite our pedigree and a very long history of successes going back to the 1870s, the majority of our new complex challenges in developing new products and services were no longer falling neatly within, and were not divisible perfectly into the standard disciplines of Mechanical, Electrical, Software, Testing, Project and People Management. One of my first major challenges as a senior staff engineer was solving the irregular dynamic performance of a gas turbine electro-mechanical fuel control valve: the *System of Interest (SoI)* which was the product we sold. The valve passed ALL the static and dynamic tests and satisfied ALL the required specs inside the company walls but would cause the gas turbine, the *Context System*, to shut down in about 5% of all locations. These locations did not have an obvious clue or common characteristic to blame. Some failures were in cold climates like Siberia, some in hot desert locations, and some in moderate or controlled climates. The shutdowns were due to the valves not tracking their commanded position in the field. When we focused on the valve position response, we saw some irregular response features but nothing that would increase the tracking error to shutdown levels. If not resolved, this issue would be a significant risk for the company because there was no way to know or identify a failure before installation on the customer gas turbines in the field.

The problem was difficult to diagnose by classical methods. All the disciplines were satisfied that they followed best practices to meet all the requirements. Furthermore, the team members were all convinced that another team member was responsible. The ME blamed the analyst for incorrectly designing and tuning the control algorithm. The analyst blamed the ME for not having estimated the correct level of friction. The electrical and software engineers teamed up to blame everyone else because although all the specs were met, the valve behaved noticeably differently in successive tests. Their solution was to improve the software and hardware to incorporate a more expensive position sensor, believing that a more expensive, accurate, and repeatable sensor would resolve all the issues. This fix made the response irregularities much worse and more accurately measurable!

Our inability to deal with the response irregularity was due to our counterproductive mindset of mainly caring about and focusing our efforts on our areas of expertise and responsibility. Through education, years of practice, a company culture that rewarded it, and the long-term effects of the way we were organized, we learned to subdivide the problem into smaller chunks that fit inside the Mechanical, Electrical, Software, Analyst, and Managerial areas of responsibility. Our dominant mental model was that if each discipline solved its chunk of the problem separately, the integration would naturally lead to a successful system.

Our team was also accustomed to linear cause and effect thinking: We tended to expect that the cause of a problem was most likely near where it was detected, in time and location. This mindset had worked well previously on more straightforward challenges. However, as the requirements became more arduous and the complexity of the products and services grew over time, the same mindset no longer helped, and we needed systems-oriented frameworks, languages, and tools to deal with the ever-increasing complexity.

The solution became more visible once we started considering the problem as a whole. Instead of the analyst designing the controller to give the maximum performance for a required level of friction determined by the mechanical engineer, we all worked together to understand all the uncertain Electrical, Mechanical, Sensor, Software, and Control interactions with each other, external components, and the operating environment. This led to a better strategy for dealing with friction: design a controller that gave the required performance over the full range of friction values the valve could encounter due to the above interactions. It turns out that optimizing or squeezing the best possible performance, as measured in a controlled lab inside the company, can become very counterproductive and can lead to very poor performance and failure in the field! Instead of aiming for the best possible performance for a narrow expected operating condition, we used a controller technology aimed at robust performance that remained graceful over a broader range of operating conditions. All our issues disappeared after adopting this better transdisciplinary mindset, and the valve product line became a success, controlling all types of liquid and gaseous fuels in the gas turbine industry at all locations, including those operating in remote locations with severe operating conditions.

Interestingly, we were initially fixated on squeezing the best possible performance with testing that appeared to confirm it. Yet, it was not until we adopted more systems-oriented thinking that we realized our efforts were making the performance in the real world worse, and we were part of the problem that caused the shutdowns.

So why would we expect a product that passes all the specs and tests inside the company walls to be naturally successful in the field? The answer is linearized compartmentalized thinking! When we are overly focused on the internal boundaries of the *SoI*, which is the product or service we are selling, we can ignore many things that are outside the scope of our responsibilities that may be critical: human factors, environmental conditions, operating scenarios, anomalies, and other systems that interact with the SoI to cause emerging behavior that is unwanted and unanticipated.

Going Back to School to Create Another

It took a Bachelor's, Master's, PhD, Postdoc, and a few years facing real-world problems in industry for me to realize that at least some of my secondary, undergraduate, and graduate education covering Aerospace Engineering, Mathematics, and Information Technology appeared to reward linear, inside-the-box, short-term, disciplinary, departmental, and analytical (divide and concur) problem-solving.

After encountering several real-world problems like the Fuel Control Valve above, it quickly became apparent that although we had all the right specialists on our team, our challenges were often not divisible into more minor problems each specialist could resolve individually. What we needed was a trans-disciplinary philosophy and rigorous methods to implement it. This motivated me to join the Systems Design and Management program at MIT in 2006, which started my journey on the systems thinking trail. In 2008, fresh back from MIT, I joined a team led by Dr. Ron Sega (x- astronaut and under-secretary of the Air Force) to help create a local Systems Engineering program at Colorado State University to develop the right kind of industry talent able to tackle complex challenges that fall somewhere in between a broad range of technical and non-technical disciplines.

To help better understand the need, we surveyed 40 companies in Northern Colorado. We identified several strong clusters of common symptoms related to complex system challenges across several industries that resembled the fuel valve problem above in many ways that were explainable with a systems-oriented mindset. The common symptoms and the opinions of respondents from these companies on how to cure them drove the initial design of the Colorado State University (CSU) Systems Engineering program. We kick-started the new program in 2009 and saw it rapidly grow into one of the nation's premium graduate Systems Engineering programs, delivering systems-oriented talent to various industries and organizations.

Systems Thinking Talent Flow into Industry

Today, despite the growing number of systems-oriented talent joining or advancing in the workforce, we see that while general awareness of systems thinking and usage of systems-oriented terminology, buzzwords, and tools have increased significantly, the dominant approach to understanding and solving complex problems remains essentially close to classical linear thinking methods. One way to measure the impact of systems thinking in an organization is to measure the usage of rigorous methods (frameworks, languages, and tools) specifically developed to enforce the application of systems thinking to complex challenges.

For example, many, including myself, believe that System Dynamics is a primary and overarching rigorous method that various technical and managerial disciplines can leverage to solve their complex challenges. Furthermore, System Dynamics can be an umbrella method to make other methods like AI/Machine Learning, Agent-Based Modeling, and Architectural modeling systems-oriented, as discussed in Chapters 10 and 11. Yet, the proportion of the workforce with practical System Dynamics skills in product and service-oriented organizations is almost negligible today.

We wrote this book to help young fresh talent of all disciplines start their systems thinking journey early in their careers without having to initially learn special theories, methods, or disciplines such as systems engineering. However, this book also provides a relatively easy path for experienced professionals who are not yet systems thinkers.

Cities, Teaching and Thinking a Journey of Systems Thinking

Steven Allen Conrad, PhD

Associate Professor of Systems Engineering (Colorado State University)

Many offerings on systems thinking today underestimate the value of thought and while many disciplines define and address specific systems well, I see—too often—a tendency to fall back into comfortable, regressive, linear thinking. I see this in my research and professional activities. This is something I encounter in teaching systems thinking to students. Moreover, I see it in the design and management of cities, where the interconnectedness of urban systems is often overlooked in favor of segmented approaches that focus on optimizing one problem at a time. These observations have shaped my journey in systems thinking, leading me to examine how to teach systems thinking effectively and how to apply it to the design and management of cities.

In 2014 I was asked to design a systems thinking class for 2nd year undergraduate students majoring in global environmental studies at Simon Fraser University. At first, I thought to introduce systems thinking as an assessment tool to evaluate global change through the application of basic systems thinking concepts such as feedback, delays, and stock and flows. A strong understanding of these concepts would surely help students in their later coursework on population dynamics, earth systems, and human influence on global issues. Recognizing and predicting system behavior would also give these students a core foundation for managing global systems. I introduced students to the six basic shapes to describe behavior: oscillations, overshoot, goal seeking, exponential growth and decay, and s-shaped growth. I had students draw a stock and flow diagram of a watershed and predict behaviors from their stock and flow diagrams of other systems. I taught students how to think visually through mind maps and causal loop diagrams and examine feedbacks through activities like the beer game simulation.

While students became proficient with the application of techniques to provided problems, I found that they still tended to sink back into linear thinking when asked to look at unique problems. So in later weeks, I added discussions of systems **thinking** principles and existing mental models that have guided students through their education. We compared these models with systems thinking views and broke down how we can apply principles that account for the unmeasurable, and that we are part of the system and are affected by the system. I had students challenge each other to find connections and change their existing mental models of the world. It was after one of these discussions that a student came to me and shared that while the course title included thinking they did not truly realize that systems thinking was about changing how we **think** about problems and the world. It was an "ah-ha" moment for them. Afterward, I updated my course to start with material on how systems thinking is a new language they will learn and that it will add new ways of thinking such as dynamic thinking, system as a cause thinking, holistic thinking, and interconnectedness as a foundation of thought. I shifted my assessments away from tests on concepts to applied principles and investigations of where systems thinking can find leverage points to change the core structure of global issues. Over the years, I have shifted my focus on systems thinking from being a tool to being a principal foundation for looking at problems and society. I find that many early systems thinkers also start with systems thinking as a toolset, but in my years teaching systems thinking, I hope that I have helped hundreds of students move away from it being a tool to viewing systems thinking as a unique language and way of thought.

Overcoming Existing Mindsets

Systems thinking should naturally apply to city water and energy systems, after all, cities are complex systems with buildings, roads, businesses, traffic, communication systems, street lighting, and countless other items that must come together to serve the people who live there. However, what I've found is that the common approach to managing cities is segmented into soiled disciplines that focus on bringing the best solutions to their specific task. The traffic engineer looks to optimize traffic flow, the building engineer looks to improve building performance, and the water engineer likewise looks to deliver the highest quality of water to citizens. When we optimize a single part of a complex system we lose sight of the complex relationships that drive cities. In efforts to provide energy to cities, we ignore the feedbacks that would see energy supplies decrease as populations grow until there is an equilibrium point where energy needs are met with energy supply. These energy balancing loops are in nature and help ensure no population exceeds their available resource supply or risk collapse. Instead, we see energy supply for cities in terms of needing excess to ensure system function. We have mental models where a city without energy to grow is a dying city. Therefore the energy manager's job under this mental model shifts from meeting the city's needs to continually injecting new energy supplies, thereby forcing growth. I have also seen brilliant city water operators focus on, what we call systems thinking, the tip of the iceberg despite managing a system that is almost entirely buried. In one case I found myself, a young junior water systems engineer, in a lively debate with a group of experienced city water professionals about how a water system was behaving.

I was tasked with developing a comprehensive water systems model for a US utility and as I began examining the system, a behavior emerged that didn't quite make sense. An area of the water system acted as though it wasn't connected to the nearby treatment plant, always requiring water from another treatment plant. When I looked at how much energy it should take to pump water to that part of the city it was also far greater than what should be needed. I suggested that there were some water distribution issues, but my suggestions were met with polite skepticism. We dismissed the behavior as a modeling quirk. After all, the operational team had years of records showing consistent water delivery to that area, so the idea of a problem with the water system seemed implausible.

But something about the behavior didn't sit right with me and I couldn't let it go. After reviewing the model and data again, I convinced the team to go into the field for a closer look. We discovered the culprit: a massive valve that should have been sending water from west to east had been long closed. Its closure forced operators to improvise over the years, developing clever but isolated solutions to compensate for the system problem.

These solutions worked, in a sense, as they kept water flowing to where it was needed. But these fixes didn't address the underlying issue. By viewing the problem through a narrow lens and treating the problem of water supply as a linear issue, it was easy to overlook the broader systemic implications. Only when we stepped back and examined the entire network, as we did when developing the system model, did the real problem reveal itself. The lesson here is that solving isolated issues without considering the whole system can mask problems and potentially create more issues than it solves. For me, that moment when the field technicians found the closed value, was a powerful reminder of why systems thinking matters—why seeing the connections, not just the components, is important.

This type of directed thinking has not been a unique experience in the 25+ years I have worked in and researched urban water and energy systems. I found dedicated problem solvers, but many who are unfamiliar with systems thinking. I found instances where water supplies are designed to reduce water pressure without consideration of capturing and using that reduction of energy. I have found energy utilities that depend on the water supply to cool power generation facilities with limited awareness or consideration of the reliability and function of that water supply. Across both sectors, I have found incredibly capable system managers who treat their systems as a series of individual events. I find no fault with this as this viewpoint only follows the regressive and structured problem-solving processes that we have learned in our education and training systems. A system that unintentionally limits the potential for innovative and integrated solutions to address the complex interdependencies between water and energy systems. A system that can however be reinvented through systems thinking.

This book was written to share these lessons about systems thinking with you. We hope that by exploring the perspectives and stories within these pages you will see how systems thinking can transform the way you approach problems.

Science and Technology Enterprise Ecosystems

Quentin Edward Saulter, MSc

Engineer (Department of Defense), PhD Candidate (Colorado State University)

I work in business and management practices pertaining to basic science research. In my 25+ years of doing so, I have not seen much change over the past few decades. In fact, the standard practice has not adapted to the ever-increasing complexity of funding and managing basic science research programs. After undertaking a PhD in Systems Engineering, concentrating on dynamic systems, I began to see where Systems Thinking would be beneficial for a Science and Technology (S&T) enterprise system.

The problem that is most encountered in an S&T enterprise system is that the benefit of funding basic science research with billions of dollars comes into question due to stakeholders' perceptions of the lack of immediate benefits from these investments. We live in an instant gratification society where time is always of the essence. A basic science research system is a complex system of solicitation, reviewing, selection, funding, and managing of basic research programs. I have been involved in all these aspects of this system and can attest to the complexity. Most S&T enterprise systems are guided by constrained inflexible waterfall processes that have not evolved in decades to keep up with the speed of new technologies, discoveries, cutting-edge research, and ideas for managing S&T that have been offered by System Dynamics and Systems Thinking domain researchers.

In an S&T ecosystem, principal investigators (PIs) in universities, government laboratories, and industry produce ideas at least every year. These researchers are not finding funding promptly to investigate their ideas in a laboratory or fund people necessary for the investigations. Also, PIs need funding that lasts for a sustained period of time to bring their ideas to fruition. Systems Thinking and System Dynamics can be used to explain the complexity of the problem, offer a solution, and show the results of dynamic simulations relating to funding, mental models of research efficacy, and management of basic science research.

Basic science research includes activities directed toward greater knowledge or understanding of the fundamental aspects of phenomena without specific applications toward processes or products in mind. It includes scientific studies and experimentation's directed toward increasing fundamental knowledge and understanding of the physical, engineering, environmental, and life sciences related to long-term national security needs. I have experience as a researcher working at a world-class particle accelerator and in supporting basic research to advance fundamental knowledge in fields important to national defense as a portfolio overseer of many research programs. I have seen concerns about whether the nature of government-funded basic research is worth the billions of dollars appropriated to fund it. Some of the concerns are from academia in which too much government research dollars are being spent for research that does not fall under the definition of basic research. Other academic and small business concerns are the S&T enterprise ecosystem is cumbersome with outdated funding and reporting requirements and lacks dedicated steady funding for good research.

Systems Thinking and System Dynamics tools could point out the issues for these concerns and identify outdated processes that hinder the effective generation of new science and technologies by incorporating the knowledge of systemic feedbacks into process thinking. Top-down or Waterfall processes may not be the most efficient tools for some aspects of generating science and technology innovations. The Waterfall process is endemic to most enterprise ecosystems for basic science research funding and execution. This sequential process follows from the overall planning, programming, budgeting, and execution of funds in an enterprise system. In some systems, it can take almost 2 years from proposing a project to funding it. In contrast, PIs from government laboratories, academia, or industry have ideas and create possible research projects in the same year they expect to be funded. Most researchers want to immediately start their research projects as soon as possible to take advantage of novel leaps in science and technology discoveries. The 2-year execution process from idea to data reduces the ability to stay on the cutting edge of research. Systems Thinking is a possible solution to this problem. The use of a mixture of Enterprise Systems Engineering and System Dynamics can be used to shorten the time frame of financial and contractual work to fund basic science research.

Systems Thinking for Engineering and Non-Engineering Cultures

Sarwat Kashmiri Chappell, MSc

Engineer (Department of Defense) and PhD Candidate (Colorado State University)

Systems Thinking has changed the way I approach problems both personally and professionally. During my career, I have had the opportunity and privilege to participate, oversee, and manage research ranging from basic science and advanced research to engineering and fielding systems. My education, training, and certifications in engineering, systems acquisition, and management did not include Systems Thinking (ST). It was not until I embarked on my journey to earn a PhD in Systems Engineering that I encountered systems thinking. Reflecting on my past experiences, I can recall many instances where linear systems thinking was applied to complex problems and how a lack of ST and ST culture affected the outcome. In one example, the Systems Engineering team had flowed down the requirements of a complex system to subsystem requirements. Per standard development practice, each subsystem team developed its subsystem specifications and based its design on them. The teams met periodically but had an inherent rivalry which is common in development teams without ST culture models. As was typical at the time, the team developing the system structure drove the overall size, weight, and power allocation of the overall system with the assumption that other subsystems would adjust and work within the allocations. It was not until the activity to integrate the navigation system within the system structure that the teams discovered that the sensor boards did not have enough space to be mounted correctly to give accurate information. It would require a redesign of the navigation system or the entire system or the use of different components in navigation which would drive cost and schedule. The two teams commenced to point fingers at each other and each team did not see themselves as part of the problem. Eventually, the entire system had to be modified, and it delayed the entire project, increased costs, and slowed down delivery from the program. This could have been avoided by applying Systems Principles such as the Holism and 3-Systems Principles to the design of the system as well as subsystems. If the team had adopted a ST culture and practiced the ways and habits of a systems thinker, many of the problems could have been avoided.

The application of Systems Thinking throughout the life cycle of technology development from basic and advanced research to engineering could revolutionize how well an idea or technology transitions to a usable capability. Research categories such as physics, mathematics, biology, chemistry, etc. have different ecosystems, i.e., silos in which the information is contained. The mental model of how research is conducted within the silo is dependent on the particular community safeguarding their information. This affects the decisions made early on in exploring an idea/technology. For example, a chemistry discovery applicable to a solvent within the chemistry silo may be used to remove rust in a system that belongs to the mechanical engineering silo. However, this may either take a long time or never be discovered since the silos do not interact. This can have long-term impacts on how a technology is adopted. This is typical of multidisciplinary complex systems and is one of the basic tenets of Systems Thinking that helps deal with the complexity of systems. For the solvent example, if it does not work well within the researchers area of expertise, it will be considered a failure in that silo and put aside. In another silo it could have been a success. Both the program managers and researchers in these communities would benefit from an ST education and an adoption of an ST culture which would enable them to incorporate different perspectives and consider the lifecycle and long-term evolutionary impacts.

A non-technical example of how systems thinking could be beneficial is in the field of health science. Literature has shown that exercise provides immediate and long-term health benefits. Benefits include improvements in movement, weight management, and strengthening of bones and muscles as well as managing diseases and reducing risk for diseases. Achieving the benefits requires sustained activity as part of a person's lifestyle. It is common for people to start exercising but stop after a short time due to a lack of motivation. This decrease in motivation to exercise often comes from the perception of too much difficulty in exercise or due to injury. Even highly motivated individuals who use exercise plans can lose motivation since the plans are fixed to a specific predetermined fitness level and do not account for the evolving individual's physical and mental state and interactions with any underlying health conditions. The plans do not consider all the interconnected physiological, environmental, and mental dynamical variables that interact with each other in a complex way to modify a person's physiology over time during sustained activity. Physiological variables include elements of a person's cardiovascular, pulmonary and muscular systems that interact continually among themselves as well as with mental state variables such as motivation, confidence, etc. in an environment that changes with temperature, terrain, etc. as a person performs an activity. These variables create dynamic reinforcing and balancing feedback loops that interact with each other over the training time so the cumulative effect on endurance is not evident till a person has trained for a while. The challenge is the cyclic nature of how one's mental state changes with the change in physical state. For example, an inability to meet physical goals leads to a decrease in confidence and, therefore, motivation, which in turn affects completing physical activity. These dynamic feedback loops between physical and mental states and their contributions to enable sustained activity has not been studied as a complex dynamical evolving system.

A system dynamical model could be used to gain an understanding of the contribution and sensitivity of physical and mental endurance factors to performance of sustained activity. Using initial data for an individual, either measured or from smart watches, the model can be customized by coaches to monitor and tailor training to an individual's changing mental and physical fitness level and minimize injury.

Seeing and Not Seeing Systems Thinking Across Industries

Jill Speece, PhD, INCOSE CSEP

Assistant Professor of Industrial and Manufacturing Engineering (California Polytechnic State University San Luis Obispo, CA)

I have worked in various industries: the defense industry, with its well-established processes; a startup manufacturing company where changes occurred daily, if not hourly; a medical device manufacturing company subject to stringent government regulations; a clinical healthcare environment, where patients themselves are unpredictable variables; and academia, where theoretical knowledge is more easily shared than practical expertise. Each of these environments stands to benefit from the widespread adoption of systems thinking as a critical skill. However, the process of fostering cultural change in each area depends heavily on its unique context. While I have seen systems thinking applied in different ways within these industries, it has not been broadly adopted across entire organizations. In the defense and medical device industries, some systems thinking tools are already embedded in existing design and review processes. In healthcare, systems thinking emerges when a team of providers collaborates to create a care plan for a patient. Despite its utility, systems thinking is not consistently practiced across organizations. Many workers, generally speaking, believe that understanding how the system operates and assessing the impact of potential changes is the responsibility of upper management or physicians.

In both industries where I am currently most active—healthcare and academia—the approach to developing systems thinking skills differs. In healthcare, specifically, integrating systems thinking into existing continuous improvement programs offers the most straightforward way to teach this skill set. Systems thinking should not be seen as competing with continuous improvement but as complementing it. The principles, approaches, and tools of systems thinking should be thoughtfully incorporated into the current continuous improvement curriculum or training. In academia for undergraduate students, one effective way to teach systems thinking is through developing cross-disciplinary teams and instructing them to use systems thinking tools to solve "wicked problems." Some of the complex problems students have tackled include eliminating the proliferation of microplastics, effectively cleaning up oil spills in the ocean, avoiding assault, and reinvigorating eroding coastlines. These are classified as wicked problems because they are complex and there are multiple potential solutions to explore.

Although the approach may differ, a common need is a resource to draw from. This text is a labor of love inspired by the recognized need for a systems thinking resource that is palatable to both undergraduate students and early-career individuals while also speaking to more seasoned professionals.

Insights on Systems Thinking from Small Tech Startups to Large Aerospace/Defense Companies

Golam M. Bokhtier, MEng, INCOSE ESEP

Systems Engineer (Northrop Grumman) and PhD Candidate (Colorado State University)

Over the past 20 years, my career has taken me through two vastly different professional worlds—the established, methodical environment of the aerospace and defense industries, and the fast-paced, ever-evolving realm of tech startups. These experiences have provided me with a unique perspective on how systems thinking plays a critical role in both sectors, though in different ways.

In the aerospace and defense industries, the journey from innovation to deployment can span a decade or more. The processes are slow and are centered on safety, reliability, and rigorous testing. Here, systems thinking is essential to ensure that complex technologies are carefully integrated and meet stringent standards before they see the light of day. Without this holistic approach, new products are often overlooked or delayed indefinitely. The weight of responsibility in aerospace means that every piece of technology, no matter how innovative, must pass through multiple layers of scrutiny before it is allowed to fly.

On the other hand, tech start-ups live by a different set of rules. The race to be first to market drives rapid innovation cycles, where companies are often focused on proving their products as quickly as possible. I have had the privilege of working with an in-flight connectivity company, where I joined during its early days with fewer than 600 people, and later with an autonomous pilotless aircraft company, which had fewer than 100 employees at the time. Both were marked by an incredible sense of urgency: teams of highly talented individuals working tirelessly to bring their visions to life. However, this rush to market came with its own challenges, especially when it came to balancing innovation with safety and long-term sustainability.

The startup environment, in particular, highlighted the critical need for a strong systems thinking culture. Although startups often aim to reduce development cycles by half, it is only with a laser focus on the entire system, the technology, the market, the user experience, and the regulatory landscape that they can hope to achieve success. Systems thinking in this context allows companies to innovate rapidly while still maintaining a sense of order and responsibility, ensuring that the product they bring to market is not only groundbreaking, but also reliable and safe.

This material in this book can drive success in both the slow and deliberate processes of aerospace and defense and the fast-moving world of tech start-ups. In both cases, systems thinking acts as a stabilizing force, ensuring that no matter how fast or slow the innovation, the final product serves its purpose effectively and safely.

Systems Thinking's Unique Value to Unknown, Unknown Problems

Martin "Trae" Span, PhD, INCOSE CSEP

Developmental Engineer (US Air Force)

In my career as a US Military officer and as an academic instructor, I've enjoyed tackling the most perplexing of problems. These problems are those in which both the cause and effects are unknown. We often have a shortsighted viewpoint of what may have caused the problem and do not fully understand the effects. These problems are the most vexing and therefore deserve the most attention. Unfortunately, most process-oriented engineering or checklist-based problem-solving solutions do not adequately address these problems. Time after time we see large organizations make decisions and spend exorbitant amounts of money on solutions that do not actually fix problems and instead cause new ones. This is unfortunately well evidenced in government and military history: Both have a great track record of solving the wrong problem or solving the right problem the wrong way leading to adverse consequences that were never intended. We see this in warfare in the Middle East where removing one bad actor creates a vacuum of power replaced with other troublesome regimes.

I always gravitated toward an exploratory mindset to understand the context around a problem before trying to solve it. Then in my academic journey, I discovered this is what Systems Thinking is about. I now proudly call myself a "systems thinker" because applying systems thinking principles is the "answer" to solving complex problems. It has allowed me to guide teams trying to gain a true understanding of the complexity of their challenges so they can create well-constructed solutions to meet user needs. Systems thinking is a key enabler in advancing a culture change in large process and checklist-driven organizations. Therefore, it is my passion.

Why Is Systems Thinking Even a Thing?

Kirk Reinholtz, PhD

Retired Principal Engineer (Jet Propulsion Lab)

My quest has forever been to gain a deeper understanding of everything I observe. One of my early questions as I began my studies of ST was why are there so many books on the subject? Why were the ST writings of the 1970s and earlier lamenting the same lack of ST we observe today? I believe it is because until quite recently there has been little value in thinking much beyond the metaphorical moment: When one meal could mean the difference between life and death, hesitating to ponder the long-term implications of the hunt would not make for a long-lived set of genes. But now, in an evolutionary instant, we have built an environment in which short-term thinking can extinguish many species, including ourselves. Grabbing in the moment will not serve us well, but our genetic programming is always nudging us to do so. This is what attracted me to ST. Though ST is classically described as a tool to improve our ability to see the cascading consequences of our actions, I believe it is more. It is a tool to improve our ability to see why we don't react to our knowledge of the consequences of our actions and consequently (hopefully) improve our ability to react, to forgo immediate reward for future benefit. This is the ST behind the ST, and a powerful reason for us all to learn and apply ST.

Fort Collins, CO, USA — Kamran Eftekhari Shahroudi
Fort Collins, CO, USA — Steven Conrad
San Luis Obispo, CA, USA — Jill Speece
Fort Collins, CO, USA — Kirk Reinholtz
Fort Collins, CO, USA — Martin "Trae" Span
Fort Collins, CO, USA — Sarwat Kashmiri Chappell
Fort Collins, CO, USA — Quentin Edward Saulter
Fort Collins, CO, USA — Golam M. Bokhtier
November 2024

Acknowledgement We acknowledge the Colorado State University Systems Engineering Department and Woodward, Inc. for their support and motivation to create the SYSE 505 Systems Thinking for the Real-World class in 2020. This class provided a rich exchange of opinions between circa 80 professionals from various industries working in a range of hard-to-soft roles. We would like to acknowledge their exchange and reflections that helped tune out disciplinary bias from the core material of this book. SYSE 505 was also the place where the authors of this book met and developed a passion for creating an easier path for young fresh talent to adopt systems thinking in their personal and professional lives.

We acknowledge Dixie Poteet (doctoral student at Colorado State University) for creating hand-drawn illustrations for the Book Cover, Several Part I, and All Chapter 5 Figures.

We acknowledge INCOSE ESEPs James R. MacDonald, PhD, and Emmet (Rusty) Eckman for their timely, detailed, constructive, and impactful review of the first draft version of this book.

We acknowledge John Metcalf, PhD Candidate at Colorado State University, for his contribution to the DSM case study and his feedback on the material in the DSM section in Chapter 12.

Declarations The opinions expressed in this textbook do not necessarily represent the opinions of Woodward, Inc., Northrop Grumman, Colorado State University, California Polytechnic State University San Luis Obispo, US Air Force, US Navy, US DoD, and US government, where the authors may have current or previous affiliations with.

This study was not funded by any academic or business organization. The authors put in their own unpaid time on a volunteer basis into the creation of all the material in this book.

Ethics Approval This textbook does not include any primary studies with humans or animals so Ethics Approvals are not applicable.

Contents

List of Figures

Acronyms

ABM	Agent-Based Model
AI	Artificial Intelligence
AI/ML	Artificial Intelligence/Machine Learning
AoA	Angle of Attack
ATC	Air Traffic Control
BDD	Block Definition Diagram
BMI	Body Mass Index
CBS	Central Base Station
CDC	Centers for Disease Control and Prevention
CDM	Conceptual Design Model
CFD	Computational Fluid Dynamics
CLD	Causal Loop Diagram
COTS	Commercial-Off-the-Shelf
CPS	Cyber-Physical Systems
CSU	Colorado State University
DDT	Dichlorodiphenyltrichloroethane
DoD	Department of Defense
DSM	Design Structure Matrix
EE	Electrical Engineering
EMA	Electromechanical Actuator
eVTOL	Electrical Vehicle Takeoff and Landing
EWQMS	Energy and Water Quality Management System
FAA	Federal Aviation Administration
FC	Flight Controls
FCAS	Flight Controls Actuation System
FEA	Finite Element Analysis
FIW	Fans-in-Wings
FR	Functional Requirements
GAN	General Adversarial Networks
GDP	Gross Domestic Product
GPA	Global Positioning System

HIPAA	Health Insurance Portability and Accountability Act
HITL	Human-in-the-Loop
ICAO	International Civil Aviation Organization
IIOT	Industrial Internet of Things
INCOSE	International Council on Systems Engineering
IoT or IOT	Internet of Things
kg	Kilograms
km	Kilometers
LED	Light-Emitting Diode
LLM	Large Language Model
LV	Logical Viewpoint
MBS	Mobile Base Station
MBSAP	Model-Based System Architecture Process
MBSE	Model-Based Systems Engineering
MBST	Model-Based Systems Thinking
MCAS	Maneuvering Characteristics Augmentation System
MCMC	Markov Chain Monte Carlo
ME	Mechanical Engineering
MFS	Management Flight Simulator
ML	Machine Learning
MSE	Mean Squared Error
MTBF	Mean Time Between Failure
OEM	Original Equipment Manufacturer
OV	Operational Viewpoint
PE	Physical Education
PM	Program Manager/Program Management
PV	Physical Viewpoint
RIS	Radiology Information System
SATCOM	Satellite Communication
SD	System Dynamics
SE	Systems Engineering
SEBoK	Systems Engineering Book of Knowledge
SFC	Surface-less Flight Controls
SMD	State Machine Diagram
SOI/SoI	System-of-Interest
SPs	Systems Principles
ST	Systems Thinking
SWs	Ways of the Systems Thinker or the Systems Ways
SW	Software/Software Engineering
SE	Systems Engineering
SysML	Systems Modeling Language
TAT	Turnaround Time
UAS	Unmanned Aircraft Systems
UAV	Unmanned Aerial Vehicle
UC	Use Case

UML	Unified Modeling Language
V&V	Verification and Validation
VAE	Variational Auto Encoders
VHF	VHF Omnidirectional Range
VOR	VHF Omnidirectional Range
VTOL	Vertical Takeoff and Landing
WF	Wildfire
WHO	World Health Organization

Part I
Why Systems Thinking

The best way to predict your future is to create it—*Peter Drucker*

Chapter 1
Why Systems Thinking?

Abstract Systems thinking allows individuals and organizations to address complex problems holistically and comprehensively. It involves understanding that various system components are interconnected and that changes in one area can ripple through others, often in unpredictable ways. By adopting a systems thinking mindset, one can better anticipate the potential consequences of actions, identify leverage points for change, and develop solutions that consider the broader context rather than just isolated parts. Systems thinking causes a shift away from a linear, cause-and-effect perspective to a more dynamic, feedback understanding of how things work. Students have been taught to follow traditional thought processes to analyze and solve problems, where a problem is broken up into manageable chunks and each piece is analyzed and solved in isolation before putting the piecewise solutions back together. This linear compartmentalized problem-solving method works for more straightforward and more routine problems. As society and technology have progressed, so has the complexity of our problems and the systems that operate in them. To meet our expanding needs, which include leveraging new technologies, the traditional approach to problem-solving appears to be creating more problems faster than we can solve them. So, we need improved ways of thinking and rigorous methods to improve our odds of success.

1.1 Getting Started

Systems thinking can be used to navigate the complexities of modern life. As we face global challenges that require collective action and shared understanding, it can offer a way forward that is both pragmatic and hopeful. Systems thinking adds valuable skills for positively impacting personal and professional endeavors. We have a societal need for adept systems thinkers, presenting an opportunity for us to be a catalyst for change. Learning to be a systems thinker is more than a *personal mastery* issue; it's a strategic move that will start individuals and organizations on a transformative journey that enables seeing and thinking about real-world problems as systems to increase our odds of success when facing complex challenges.

K. Eftekhari Shahroudi et al., *Practical Systems Thinking*,
https://doi.org/10.1007/978-3-031-89394-0_1

We begin this chapter by explaining some of the key terms and concepts that are fundamental to systems thinking to make it easier to understand the motivation for systems thinking in Part I, systems thinking philosophy in Part II, and particularly the *Systems Principles* in Chapter 5 that is the central theme of this book. We use simple examples and short stories that highlight the concepts and show how applying a linear problem-solving approach to complex problems can cause new or worse problems and how changing the approach can lead to better problem-solving.

We intend Part I of this book to incrementally refine our systems concepts and terminology so that we can better learn Parts II and III. We encourage the first-time reader or those new to systems thinking (or systems-oriented) terminology to casually scan the material of this chapter and try to get an overall feel of systems thinking concepts versus a very accurate understanding at this stage. For example, the Tom/Jerry and Waymo sections at the tail of this chapter are just an excuse for early exposure to the *Systems Principles (SPs)*. We do not need an accurate understanding of SPs until Chapter 5.

1.2 What Is a *Real-World* Problem?

Real-world problems are practical issues, obstacles, or behaviors that are not readily understandable or explainable by a narrow disciplinary focus or simple linear cause-and-effect relationships. They do not fit purely inside the disciplinary principles of management, economics, physics, chemistry, biology, mechanical engineering, electrical engineering, software engineering, social sciences, mathematics, etc.

A key point is that just because we decided to divide our universities into many narrow specialties, it does not mean that our real-world problems will be perfectly divisible in a controlled fashion precisely into those specialties. The issue is not that there are different disciplines. The issue is that for many real-world problems, it is the synergy of several disciplines that may lead to better understanding and identifying the points of leverage. Simply working in silos and adding isolated results and understanding individual disciplines are not good enough when the degree of complexity is high.

This explains why people with advanced degrees in narrow but deep specialties initially tend to have difficulty adjusting when they join an industry where the problem requires synergy between the disciplines and not a simple addition of the results that were produced in isolation by each discipline.

1.3 What Is a *System*?

According to the Systems Engineering Body of Knowledge (SEBoK [1]), a system is a set of interacting elements organized to achieve a specific purpose. Using this definition, if there is no purpose or interaction between the elements, then just a collection or set of elements is not considered a system. A system can be purely physical, like a house, train, or plane. It can be virtual, like a computer game, or sociotechnical, like the US Social Security System. The concept of *systems* is a big deal because it allows us to view our real-world problems as systems, thereby making them easier to understand and deal with.

We define a *system* as a *set* of *interconnected* elements that together form a *structure* and exhibit *behavior* that fulfill a *purpose* as a *whole*.

The *elements* are system components, each with characteristics that may be unique. The *elements* do not operate in isolation. They depend on each other and are *interconnected* to form a *structure* that exhibits a *behavior* as a whole. Here, the behavior or state of one element can affect others. The interconnections of elements can be physical, such as wires in a circuit, or conceptual, like the relationships between departments in a company. The *purpose* of the system is the overall objective or reason for its existence. The *purpose*, *structure*, and *behavior* of a system are not independent but they *influence* each other. For example, a metal ladder, metal steps, and bars can be joined together to create a ladder that works well and is reasonably safe to use to gain a height of up to 15 feet. However, if we significantly increase the height and safety requirements, the same design will no longer satisfy the required purpose. In this case, we would need to join the metal components differently to create a new structure that behaves in the desired new way. Changing any one of the *purpose*, *structure*, and *behavior* causes changes in or influences the other two. For example, changing the *purpose* requires changing the *structure* and *patterns of behavior*. Figure 5.3 shows an example of how the *structure* and *behavior* changes to fulfill a different *purpose*.

If we are not already familiar with a system, the easiest one of the three concepts to see, touch, feel, or understand is typically the accessible portion of the *patterns of behavior*. These are the regular, recognizable ways a system appears, operates, moves, or changes over time. The most challenging portion of an unknown system to figure out is typically the *purpose*. For example, suppose an alien civilization with no familiarity with Earth and human civilization finds Elon Musk's fully charged and unlocked electric roadster in the heliocentric orbit around the sun[1] far from Earth. They will likely be excited to find such a fascinating artifact! But what is it? The roadster will be confusing because it does not appear to have any means of propulsion, navigation, or protection for a transport device operating in space. They will probably take it back to their home planet for further study. By playing with the various parts and controls, the first thing they will probably figure out is how the mystery device behaves, e.g., how the steering wheel motion is connected to the wheel and, eventually, how the roadster moves and turns. So they will figure out the *patterns of behavior* first. The *structure* is more challenging to figure out: this is the design of components and how they are interconnected to create the *patterns of behavior* that they figured out earlier. We can probably all agree that

[1] See the wiki page for the orbiting Tesla https://en.wikipedia.org/wiki/Elon_Musk%27s_Tesla_Roadster.

the most challenging part to figure out is the *purpose* of the mystery artifact: The stated purpose of the roadster was to demonstrate that the Falcon Heavy rocket could deliver a payload from Earth to Mars. The normal earthbound purpose of the roadster, which is to carry humans and cargo while having fun, would be hard to figure out for the aliens since they do not see it working in its typical operating environment.

The method by which we try to discover, deduce, or understand the *structure* and *purpose* of a system from the *patterns of behavior* is called *systemic learning* as illustrated in Fig. 5.14.

1.4 What Is *Emergence*?

In this book, we are primarily interested in systems that are *more* or *much more* than the sum of their *parts*. We also define the main *parts* of such a system: *elements*, *interconnects*, and *purpose*. The system that we are responsible for understanding, creating, developing, improving, buying, selling, or maintaining is called the *system of interest SOI*. Systems literature often cite the *Holism Principle*:

Holism Principle (Complex) *Systems* are (much) *more* than the *sum* of their parts—*elements*, *interconnects*, and *purpose*.

But how can anything be more or much more than the sum of its parts? The answer is *emergence*. This is where the characteristics of the whole system are not directly attributable to the characteristics of the elements. For example, probably the most important element of a bee colony is the queen bee. It is the biggest and most powerful bee in the colony that interacts and collaborates with other bees in the hive to ensure the survival and growth of the colony. Let's call the bee colony the *SoI*. The *SoI* interacts with other systems in its environment. One major benefit that *emerges* from the interaction of bees with their environment is the fertilization of plants and trees. The fruit that results from this beneficial *emergence* does not have any characteristics that are simply or directly related to the characteristics of the bees in the colony. Even the most important bee, the queen bee, cannot point to any features in the fruit, such as size, shape, weight, smell, color, design, etcetera, to claim ownership!

The bee colony goes about its own business of survival and growth. Yet in the process of doing so, it has an immense influence on the ability of plants to procreate and bear fruit. This, in turn, impacts the survival of other beings and the growth of new plants that now have feedback influence on the bees themselves. Shifting our perspective to inside the hive, we see that the colony itself has recognizable *emergent* patterns of behavior that result from the collaboration of the different types of bees with each other. As a colony, they survive and procreate, something that the different bee types, including the queen, cannot achieve in isolation. So the bees' health, growth, migration, survival, and influence on other creatures in their ecosystem are all examples of *emergence*.

1.5 What Are the *Three Systems*?

Consider the US women's national soccer team competing in the Paris Olympics in 2024 who won gold as the *SOI* where we are trying to understand how this recently formed young team managed to win the gold medal against teams that had more seasoned star players. This system has elements (11 players on the field, a coach, an assistant coach, reserve players, etc.) and interconnects (how they are organized on the field as forward, mid, and defense and how they collaborate or perform on the field using various strategies in their playbook).

Often, to understand the success or failure of the SOI, we need to consider at least two other systems, the *system context* and the *enabling system*, as explained in detail in Chapter 5. The *system context* is the environment or the system that the *SOI* fits into. In this case, the soccer field, stadium, and the whole Olympic tournament can be chosen as the system context depending on the exact issue we are trying to understand. The referees, soccer ball, field lines, soccer rules, and athlete transportation to the stadiums can all be considered *enabling systems* also depending on the issue at hand. Chapter 5 will teach us that the best choice of the exact boundaries of the *SOI*, *system context*, and *enabling system* is very impactful and depends on our perspective, roles, and responsibilities. Systems thinking encourages us to balance our efforts and focus between the *three systems (SOI, enabling, and context)* in order to increase our chances of understanding complex real-world issues and finding the best leverage on them.

The YouTube video in Fig. 1.1 discusses the reasons behind the US women's soccer team's success. When we try to understand something, we often look for a very simple, narrow, or linear explanation, such as allocating the win to star players. In this case, it is definitely not possible to argue that the individual members of the team were fitter, more experienced, had more previous wins, or wore better boots than the teams they knocked out in the tournament. But one thing was sure: they repeatedly managed to overcome opposing teams until the final gold victory because they were consistently a better *team*. The *team* effect here is an example of how a system is *more* than the sum of its parts.

Sometimes, it is not possible to allocate a win in soccer to better teamwork like the above example. There are a few famous soccer champions who manipulate the rules and referees to get awarded disproportionately higher penalties than others. So, acting and making an interaction look like a foul in the penalty box from the viewpoint of the referee are now one of the skills that a few players have added to their athletic abilities. Without passing judgment, we can see that the advantage of the trick is not coming from the elements inside the *SOI* (i.e., the team members) but the interaction with the rules, referees, and defending players on the opposing team that are outside the *SOI*. This is an example where focusing purely inside the *SOI* will not lead to better understanding.

The concept of *three systems* has value here because it encourages us to look for *emergence* that results from their interactions. The benefit of considering the *three systems* is further elaborated and discussed in Chapter 5, which also gives a more

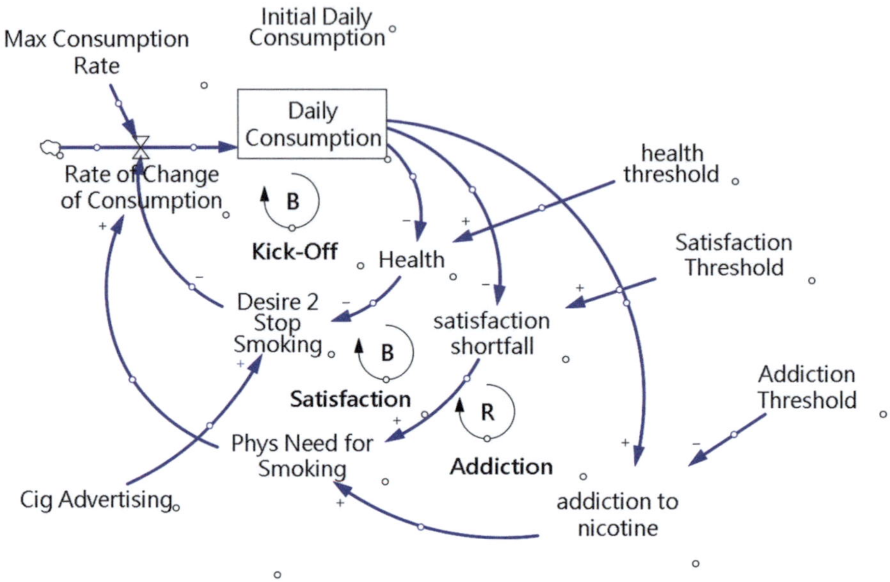

Fig. 1.1 The US women's national soccer team Olympic gold victory shows that the *system* is *more* than the sum of its *parts*

detailed discussion of how a system can be *more* or *much more* than some of its parts with more examples.

1.6 What Is a *System of Systems*?

A system of systems is a group of systems that are significantly independent of each other, but they collaborate, compete, or integrate to achieve a greater common purpose. They may have separate ownership, leadership, operators, users, stakeholders, resources, and purposes. Yet, what emerges from their collaboration is a behavior aligned towards a definable purpose. For example, the three US government branches, legislative, executive, and judicial, have separate leadership, funding, rules, processes, personnel, and defined individual purposes. These are make laws, execute laws, and evaluate laws, respectively. They may collaborate or even compete to attain their goals. Yet, balancing the powers between them and their integration is arguably the most successful form of people-led government in history. This success would not have been attainable by any of the three branches acting alone.

SOS systems thinking involves ensuring that each system fulfills its purpose. It also requires ensuring that unique capabilities or emerging beneficial behavior results from their collaboration, competition, coordination, and integration. Systems thinking gives us the right principles, behavior, and tools to manage the complexity of integrating multiple systems of interest, enabling systems, and context systems. This way of thinking has proven substantial benefits in space exploration, military, defense, and industrial processes, where independent systems must collaborate to accomplish some common purpose.

1.7 What Makes a System Complex?

The concept of complexity goes hand in hand with a system. A complex system is a system that evolves, whose behavior can change over time, and which can exhibit emergent behavior. Complexity in systems refers to the degree of interconnection and interdependence among the system's elements. A complex system is characterized by a large number of elements that interact in non-simple ways, leading to behaviors that are not easily predictable. Emergent behavior refers to the phenomenon where the overall system exhibits behaviors and properties that are not apparent or predictable from the behaviors and properties of the individual elements. These behaviors emerge from the interactions among the system's elements and are often surprising and counterintuitive. For example, in a flock of birds, each bird follows simple rules like maintaining a certain distance from its neighbors and aligning its direction with nearby birds. But the flock as a whole can exhibit complex behaviors, like forming patterns and shapes or swiftly changing direction, which are not encoded in the rules followed by individual birds.

These are examples of emergent behaviors. So, the complexity of systems is often due to emergent behavior because the interactions among the system's elements and within the three systems can lead to unexpected and complex behaviors at the system level. Understanding this concept is crucial for managing and working with complex systems, as it helps us anticipate and respond to the system's behavior, even when it's not easily predictable from the behavior of its parts. It can enable identification of the key leverage points to effect a change in the behavior of a system.

The human aspect of systems plays an important role in the behavior of systems and contributes to the complexity of a system. Humans have diverse beliefs, perspectives, and understandings of systems and of systemic interactions which leads to complexity during system development or solving a complex system problem. Human behavior is an integral part of any complex system whether it is an enterprise, environmental system, or a technological system. Mental models are the perspectives and frameworks that humans use to interpret the world. They are cognitive models that each person uses based on all their experiences and knowledge. A person will have different mental models for each situation. For example, a person who has lived in a country and never travelled will have a mental model of another country that will be fundamentally different than a world traveler. A person's socioeconomic status, gender, ethnicity, age, and life experiences can affect their mental model. Often, multiple people will adopt a similar mental model based on a global standard such as the SI system of units or agreed-upon treaties or accords. Chapter 5 shows how mental models are the highest point of leverage to effect a change in systems behavior. Mental models can be changed as humans learn a new perspective and adapt to new experiences and knowledge. The human ability to learn and its direct impact on systems give us a key leverage point to determine how to change a system behavior to solve a problem.

1.8 What Is a System *Life Cycle* and *Evolution*?

Systems have a life cycle and evolve over time. A system's life cycle refers to the series of stages that a system goes through from its inception to its retirement or death. The concept of evolution recognizes that systems are not static; they change and adapt over time. Systems evolve in response to changes in their environment, the needs of their users, technological advancements, and many other factors. This evolution can involve changes to the system's elements, interconnections, behavior, and even its purpose. Understanding the life cycle and evolution of systems is crucial for effective system management and improvement. It allows us to anticipate and plan for changes, make timely and informed decisions, and ensure that the system continues to fulfill its purpose effectively throughout its life cycle. It's a key aspect of a holistic and dynamic approach to understanding and working with systems.

For the bee system example, the worker bees perform different tasks in synergy at different times in their life cycle such as housekeeping; feeding the queen, drones, and larvae; collecting the pollen and nectar; and making the wax. In the first 15 days, they work together inside the hive as "house" bees, and in the second half of life, they work outside the hive as "field" bees. They sometimes exhibit emergent behavior where they kill and replace the queen bee with a more competent one. Emergent behavior is the complex, often unpredictable, behavior that arises from the interactions among a system's elements. This behavior can be either wanted (beneficial or desired) or unwanted (harmful or undesired). Feedback loops are a crucial mechanism through which emergent behavior arises. A feedback loop occurs when the output of a system is used as its input. There are two types of feedback loops: Positive feedback loops amplify changes and drive the system away from its initial state. They can lead to exponential growth or decline and often result in unstable, nonlinear behavior. Negative feedback loops counteract changes and help the system maintain stability. They tend to dampen fluctuations and return the system to a desired state or equilibrium. The delays in feedback loops refer to the time lag between an action and its effect. Delays can significantly influence the behavior of a system, often making it more complex and harder to predict. Chapters 11 and 12 introduce methods and tools to apply systems thinking to analyze and solve real-world complex problems. The next section about Tom and Jerry highlights the need for systems thinking and shows that complex systemic problems cannot be solved with single-point solutions.

1.9 What Is a *Mental Model*?

A mental model is a human perception, mental representation, or abstraction of the real world with a purpose to understand and influence. So mental models do not just impact the way we think. They also impact the way we interpret the real world. This influences our actions, behaviors, or the solutions that we create.

For example, the policies that we create to improve the homelessness problem in the USA are heavily influenced by our mental models about the causes. If our mental model is very narrow and linear, we may conclude that homelessness is only a function of the shortage of living space. This will lead to government policies that will only reward the creation and occupation of shelter beds without addressing all other issues that have a significant influence. Peter Stroh [8, 9] shows how this well-meaning policy was implemented and worsened the problem over time in many locations while including examples of better mental models that have led to better more sustainable improvements in homelessness in specific areas.

Chapter 3 discusses several real-world problems (childhood obesity, homelessness, and the 737 MAX) showing how narrow, linear, or compartmentalized mental models lead to counterproductive actions, behaviors, decisions, and solutions.

Systems thinking adds a philosophy and method to creating and developing mental models, making them more reliable when understanding real-world problems and leading to sustainable ways to leverage them.

1.10 What Is *Systems Thinking*?

Like the term *complexity*, understanding what systems thinking is and how to define it is evolving. There is not a universal understanding or agreement about systems thinking as there are multiple schools of thought. Yet there are at least a dozen excellent definitions that all help highlight one or more of the many facets, subtleties, quirks, and features of systems thinking and what it enables.

Below are a few definitions or explanations to highlight some of the many facets of systems thinking:

- Jamshid Gharajedaghi (*Systems Thinking: Managing Chaos and Complexity* [3]): Systems thinking is the art of simplifying complexity. It is about seeing through chaos, managing interdependency, and understanding choice. We see the world as increasingly more complex and chaotic because we use inadequate concepts to explain it. When we understand something, we no longer see it as chaotic or complex.
- Alliance for Health Policy and Systems Research [4]: Systems thinking is an approach to problem-solving that views problems as part of a wider dynamic system. It recognizes and prioritizes the understanding of linkages, relationships, interactions, and interdependencies among the components of a system that give

rise to the system's observed behavior. Systems thinking is a philosophical frame, and it can also be considered a method with its own tools.

- Peter Senge (*The Fifth Discipline Fieldbook: Strategies and Tools for Building a Learning Organization* [6]): Systems thinking [is] a way of thinking about, and a language for describing and understanding, the forces and interrelationships that shape the behavior of systems. This discipline helps us to see how to change systems more effectively, and to act more in tune with the natural processes of the natural and economic world.
- Peter Senge (*The Fifth Discipline: The Art and Practice of Learning Organization* [5]): Systems thinking is a discipline for seeing wholes. It is a framework for seeing interrelationships rather than things, for seeing patterns of change rather than static "snapshots." It is a set of general principles—distilled over the course of the twentieth century, spanning fields as diverse as the physical and social sciences, engineering, and management. It is also a set of specific tools and techniques, originating in two threads: in "feedback" concepts of cybernetics and in "servomechanism" engineering theory dating back to the nineteenth century.
- John Sterman (*Business Dynamics, Systems Thinking and Modeling for a Complex World* [7]): Many advocate the development of *systems thinking*- the ability to see the world as a complex system in which we understand that "you can't just do one thing" and that "everything is connected to everything else". If people had a holistic worldview, it is argued, they would act in consonance with the long-term best interests of the system as a whole. However learning about complex systems when you live in them is difficult. We are all passengers on an aircraft we must not only fly but redesign in flight.
- Michael C. Jackson (Keynote Presentation at INCOSE IS 2024): What is systems thinking? A multi-perspective mindset: mechanistic, interrelationships, organismic, purposeful, societal/environmental.

It is not the intent or scope of this textbook to force the reader to read a huge number of references or rank dozens of definitions of systems thinking. Our goal here is to give the fresh young talent a good starting definition and explanations that do not have a particular management, engineering, or scientific spin, but very useful and practical regardless. For example, we do not find it helpful at this stage to get into precise granular definitions of complexity, particularly since there is a significant diversity of thought as to what complexity exactly means and how to categorize or measure it in a way that is practically useful to most. For example, in this book, we do not worry if some experts call the 737 MAX problems a *mechanistic*, *cognitive*, *epistemological*, *complex*, or *just complicated* problem! What we really care about is that the real-world problems with the 737 MAX are better understood and influenced such that passengers can travel reliably, safely, cost-effectively, and comfortably. Kids learn to speak by doing and not first learning a heavy book of grammar with in-depth understanding of the ancient root of words!

By analogy, we feel that fresh young talent will more naturally pick up and do systems thinking by learning to apply a few practical *Systems Principles* versus diving into strict definitions and categorizations of complexity. Naturally, grammar is essential after kids learn to communicate first. By analogy, we recommend that fresh young talent continues their systems thinking growth after this introductory book. Chapters 5 and 7 include additional references for further learning.

Some key explanations that the majority of systems thinking literature appears to agree on is that systems thinking:

- Is about understanding difficult-to-understand (and therefore predict) *real-world* (complex) problems.
- Recognizes that the degree of *complexity* depends on *perspectives*, *methods*, and *mental models*.
- Recognizes that *real-world* problems do not necessarily fall inside pre-defined *disciplinary boundaries*.
- Requires *philosophy* but also requires *doing*.
- Uses *rigorous methods* (frameworks, languages, and tools) that are not limited to specific disciplines or domains.
- Considers how the mental models, purpose, parts, interconnects, and behavior of system*s* *influence* each other.
- Does not prioritize *quick solutions*: Instead, it encourages us to predict the *unwanted consequences* of our actions and solutions.
- Encourages a *balance* between *outward*, *inward*, and other views of systems to find the most useful *perspectives*.
- Finds *leverage* on real-world problems so they are easier to *understand* and *influence*.
- *Philosophy* and *methods* are based on *Systems Principles* that are always true and helpful in the context of *real-world* (complex) problems.

So, finally, here is a simple "helicopter" definition that tries to merge all the above key points about systems thinking:

Systems thinking is an inter-relational transdisciplinary*philosophy*, enforced by rigorous *methods*, for *understanding* real-world *complex* problems and the *systems* that may *influence* them.

1.11 How Is *Systems Thinking* Different?

This section explains the difference between *systems thinking* and the classical *linear compartmentalized thinking*.

Variants of the fishbone diagram are very common in the industry for capturing, understanding, and communicating the causes of some desired outcome or some problem that needs to be understood and resolved. Figure 1.2 is a simple example of this diagram trying to capture the causes of daily consumption of cigarettes. It is not meant to be exhaustive: for example, there are many causes not captured in this diagram. Only four major causes, factors, groups, or branches of causes are shown that drive daily consumption. Fishbone diagrams can be very useful, but most people who use the fishbone diagram exhibit a linear compartmentalized mental model of

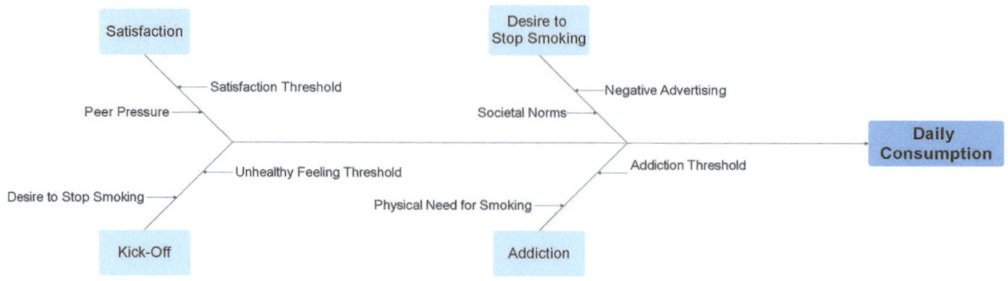

Fig. 1.2 A simple fishbone diagram of several causes of daily consumption of cigarettes

the problem. The fishbone does not offer any help to improve the limited mental model because of the following:

- **Compartmentalized**: The typical assumption is that each branch of the fishbone is independent of the other branches. This implies that you can solve the daily consumption problem by focusing on satisfaction, kickoff, desire to stop smoking, and addiction separately. The problem is that, in reality, these branches are interrelated and their interplay gives rise to complex behavior that is not directly predictable by summing the effect of the four branches.
- **Linear**: The typical assumption is that we just have chains of feed-forward effects and there is no feedback. Using a fishbone diagram, you would not be naturally encouraged to discover how daily consumption influences addiction or any other branch. This idea that cause and effect work in only one direction is counterproductive when we try to understand real-world problems that are full of circular or looping influences.
- **Missing Signs or Causal Direction**: Often, there is no encouragement to identify the sign or direction of the cause-and-effect relationships in a fishbone diagram. For example, in Fig. 1.2, it is not obvious whether a change in "satisfaction" causes "daily consumption" to change in the same or opposite direction.

By comparison, Fig. 1.3 shows how systems thinking is different because it tries to understand the interplay of causal relationships including when they form loops to *reinforce* or *balance changes*. For example, you can see that a positive change in "daily consumption" drives more addiction and "physical need for smoking" leading to further reinforcing the "daily consumption" in the same direction. Hence, the addiction loop is identified as a *reinforcing casual loop (R-Loop)*.

The figure also shows the "kickoff" and "satisfaction" loops are *balancing causal loops (B-Loops)* because a change in "daily consumption" will result in these loops trying to fight and balance that change back onto "daily consumption". So, for example, "daily consumption" may increase at first because of the effect of the "addiction" R-Loop, but then "kickoff" and "satisfaction" B-Loops can kick in to fight or balance that growth.

Comparing the classical fishbone of Fig. 1.2 with the system dynamics model of Fig. 1.3 is a good way of seeing how systems thinking is different from the

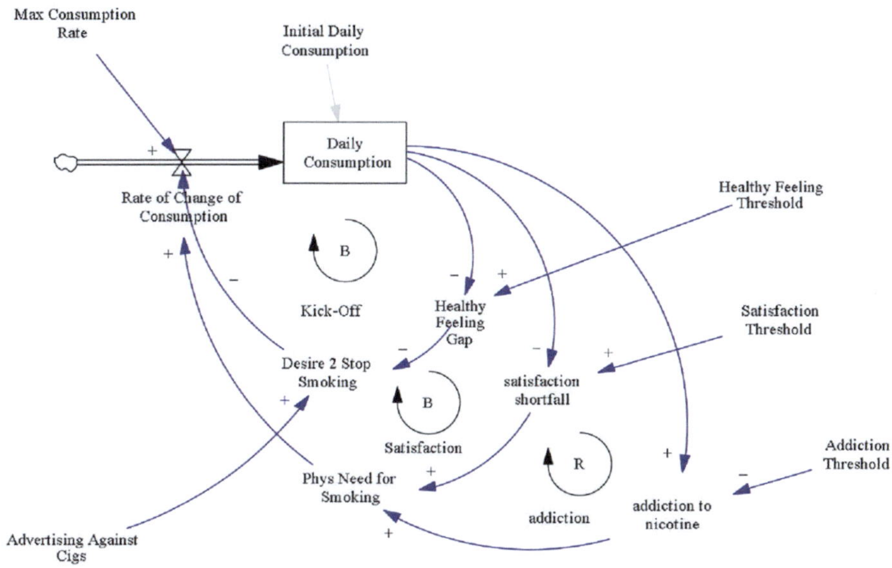

Fig. 1.3 A system dynamics view of cigarette addiction

classical linear compartmentalized thinking: first note that in systems thinking all the causal arrows require a sign to indicate the direction of the cause-and-effect relationship. These signs are necessary because in systems thinking we are no longer just interested in isolated or stand-alone linear cause and effect but we are interested in how they combine to drive emergent behavior. Systems thinking considers *interactions*, *reinforcing*, and *balancing* effects to make it easier to understand real-world behaviors. Why is this a better way of understanding real-world problems? Because wanted or unwanted real-world emergent behavior is often due to the interaction of feedback loops with delays within and between the *system of interest*, *system context*, and *enabling systems*.

For another example, Fig. 1.4 shows the farm-to-consumer process begins with farms getting money to run their farms. Farmers then cultivate crops or raise livestock on a farm. Once harvested or matured, the food is then processed and packaged. Following this, the goods are transported to markets or retail outlets, where consumers can purchase fresh farm products, either directly from stores or restaurants. This linear process is a simplified connection for consumers to have fresh quality foods available at any time. However, the linear view misses how even the downstream parts impact the upstream parts of the food supply. To understand the complexity, we need a systems-oriented causal view that captures reinforcing and balancing circular cause and effect such as that shown in the lower part of Fig. 1.4. In a complex farm-to-consumer cycle, banks play a pivotal role by providing loans for farmers to cultivate crops. Once harvested, these crops are sold to grocery stores, to restaurants, or directly to food delivery services,

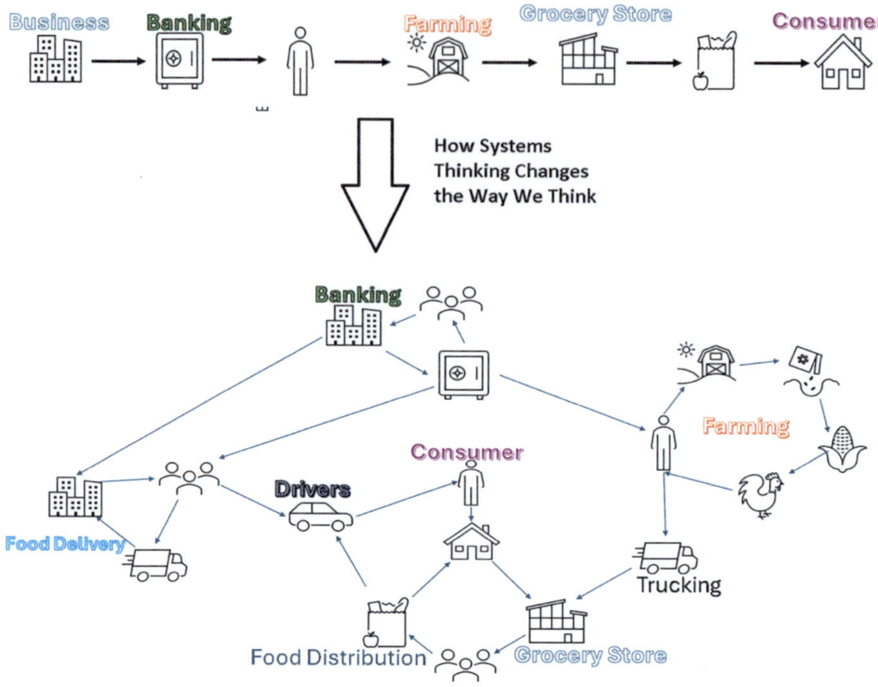

Fig. 1.4 A linear compartmentalized view vs. systems thinking view of the food supply chain

which then distribute food to consumers. The revenue from sales is deposited back into the banks from farmers and grocery distribution, ensuring a continuous cycle of investment, growth, and consumption. This process sustains the agricultural economy and ensures that consumers have access to fresh foods through a network of interconnected services. A systems thinking view of the process enables looking at the life cycle of a system and its evolution.

1.12 What Is a *Rigorous Method for Systems Thinking*?

A method is rigorous if it systemically (i.e., by design) and systematically (i.e., by a framework, process, or machine), reliably, and repeatably enforces the disciplined application of principles, laws, or knowledge for some identified purpose. Two prevalent categories of purpose are *understanding* a problem or challenge and creating or finding *solutions* for them. A rigorous method typically contains a framework or process, language(s), and tool(s) that enable synthesis, analysis, and testing to verify and validate the characteristics of our *understanding* and *solutions*.

For example, the scientific method has reliably helped the progress of science for centuries. On January 16, 2025, the Google Search Labs defined the scientific method in this way: "The scientific method is a systematic process of gaining knowledge through observation, asking a question, forming a hypothesis, conducting an experiment to test the hypothesis, analyzing the data collected, and drawing a conclusion based on the results, allowing for the objective establishment of facts through testing and experimentation". This is a good description, but it does not explicitly mention that the described process can be repeated so that a problem, question, or hypothesis can be iteratively updated to improve an understanding or a helpful solution incrementally. So, the method works like a machine, engine, or clock that keeps turning and ticking. The history of science shows that the scientific method has reliably resulted in progress that is at least piecewise stable and even partially contributing to and compatible with breakthrough discoveries and innovations that are abrupt and not incremental.

We typically need languages and tools that make accomplishing the steps of the framework of the method possible and easier. For example, Isaac Newton is accredited with developing *calculus* to describe and understand the motion of comets and planets. So calculus, which includes differential equations, is the particular *language* that made it easier to express, capture, and understand dynamic problems. Initially, differential equations were solved by hand or a slide rule. Today, we have computer software (like Mathematica or Simulink) to solve differential equations. In this case, the slide rule or Mathematica is a *tool* that makes the framework's synthesis, analysis, and testing steps easier.

We introduce *rigorous systems thinking methods* in Part III of this book. A *rigorous systems thinking method* is a method that reliably and repeatably enforces the application of *Systems Principles* (defined in Chapter 5) for understanding and finding solutions to real-world complex problems. Chapter 11 introduces the model-based systems thinking (MBST) method, which is a rigorous systems thinking method. MBST is an expanded version of the system dynamics method [2, 7]. The expansion allows the inclusion and alignment of additional modeling languages and methods like architectural, AI/machine learning, and agent-based modeling that have proven helpful in understanding and dealing with complex problems.

1.13 Tom and Jerry Trap Happy Episode

Tom and Jerry's cartoons are fun because Jerry, much smaller and seemingly weaker than Tom, often manages to escape Tom's trap. Tom always imagines the challenge of capturing Jerry to be simple. He typically picks a quick solution while ensuring that he will trap Jerry this time. Yet capturing Jerry always proves to be anything but simple! Jerry usually manages to find a way to escape! Figure 1.5 is a scene from the Trap Happy episode where Jerry pushes down on a high leverage point of the dustpan so Tom is not able to lift it until Jerry lets go of the dustpan, so it hits Tom's face very hard.

Fig. 1.5 Tom and Jerry Trap Happy episode: Jerry typically manages to find the highest point of leverage so his little efforts are significantly amplified

Fig. 1.6 Jerry's better perspective and mental model of the problem always tilts the outcome in his favor

In the Trap Happy episode, many of Tom's quick and simple solutions fail, so he calls in a professional mouse trapper. The mouse trapper is well equipped and helps Tom gas Jerry's hole. Tom and the trapper are so confident in their attempt, only focusing their efforts on directing the gas plume that they do not see Jerry coming out of the hole in plain sight wearing a gas mask! Figure 1.6 shows how surprised they are to see Jerry comfortably standing out of harm's way and enjoying the show. In the end, Jerry's better mental model prevails allowing him always to stay one step ahead of attempts to capture him. The episode ends with the trapper getting so frustrated that he decides to change his profession to a cat trapper and starts hunting Tom!

Fig. 1.7 Picking a quick solution and not worrying about the consequences of our actions, or actions by others typically backfire when the problem is complex

A subtle aspect of the story is that the skilled expert trapper that Tom hires fails just as spectacularly as Tom did, not because the trapper has bad equipment or tools or experience but because of the limited perspective and bad mental model of the complexity of the challenge in Tom and the trapper's heads. There are, unfortunately, many examples of real-world (spectacular) failures that happen to very well-certified, educated, experienced, and reputable individuals and organizations. Just like the trapper, the problem is not with their specialist knowledge or tools. The problem is with not balancing their efforts and focus into understanding the complexities of the challenge versus running with their quick solutions....

In Jerry's Gopher Girlfriend episode, using a narrow and limited mental model of the situation, Tom tries to keep Jerry and his girlfriend from stealing from a vegetable garden. After a few failed attempts, Tom tries a more severe solution: he puts several dynamite sticks in the vegetable beds. While he is waiting for the dynamite to go off, Jerry and his girlfriend move the dynamite and put it back at Tom's feet! Figure 1.7 shows Tom's confident moment just before the dynamite goes off! Pure focus on his solution and a limited mental model that does not worry about what is happening outside the scope of his solution are to blame. Tom does not balance his efforts and focus on predicting what Jerry's response might be, whether the dynamite can harm the vegetables in the garden or whether his dynamite can be repurposed to blow up in his face!

While these cartoons are made-up stories and hilarious to watch, it is unfortunately a frequently observable occurrence where well-meaning policies, efforts, or solutions by educated, experienced, and reputable organizations end up creating significantly larger problems for themselves and others. When problems get worse despite significant efforts and expenditures to improve them, then a likely cause is the lack of appreciation of the complexity of the problem or solution or both. Lack of systems thinking at the individual or organizational and cultural level is typically the ultimate culprit.

Chapter 5 gives the *Systems Principles* that we can use to improve our odds of success as we face complex problems. However, before we can properly understand and appreciate their deeper meaning and implications, we need a better understanding of what systems thinking is in practice and learn some essential systems thinking terminology and concepts that Part I provides.

1.14 Humans Impacting Autonomous Taxis in San Francisco, California

The Cruise and Waymo rollout in California is an interesting story of a well-meaning effort that lead to unanticipated and unwanted complex emergence. The idea is that autonomous taxis (also called Robotaxis) would be very cost-effective, convenient, efficient (i.e., good for the environment) and super safe as humans are generally considered to be the leading cause of accidents. In addition, the independence of people with disabilities would be better served. Soon after the rollout, a number of setbacks were reported by Forbes:

- Causing Accidents: The San Francisco Fire Department reported 55 incidents in 2023 where self-driving cars posed a problem.
- Shutting Down: Refusing to move when encountering an unexpected problem like "downed wires" or "dense fog."

It may be reasonable to expect problems in operating scenarios falling outside the data on which the Robotaxis were trained. Over time, several of these can be identified and mitigated over time as the new data is integrated into the training and tested to ensure proper response. Yet a series of Forbes articles highlighted several more complex emerging problems as shown in Fig. 1.8:

- Increasing safety concerns. Some activist groups fearing that the city is putting profits and politics ahead of peoples' safety (and driver livelihood) would put traffic cones on the hood of Robotaxis causing them to shut down requiring a service engineer to reboot the software!
- T-shirts with red stop signs becoming a hot seller because of their effect to cause the Robotaxis to stop!
- Increased congestion due to people jumping or crossing in front of easy-to-recognize Robotaxis, regardless of traffic signals, since they were assured that the Robotaxis always stopped!
- Drug use and distribution since state laws did not apply to passengers in a "bus, taxicab or limousine for hire." In addition, Waymo and Cruise policies against drug and alcohol use were not initially an effective deterrent!
- People admitting to reporters about having sex in the back of Robotaxis.

Waymo (from Google) and Cruise (from GM) were the result of roughly a decade of focused innovation and implementation to create fully autonomous self-driving taxis. Yet we see that well-meaning efforts to roll this out lead to novel emergent complex behaviors that were hard to predict early on at the design or implementation

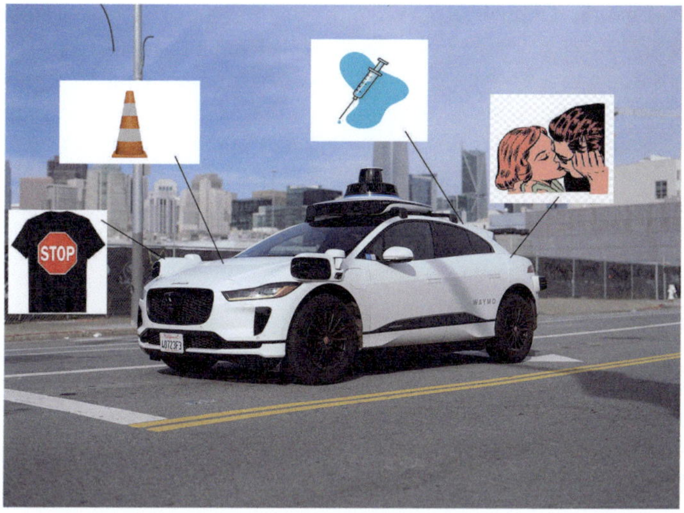

Fig. 1.8 Rollout of autonomous taxis in San Francisco leads to unanticipated and unwanted emergent complex behavior! Original photo from Wikipedia: *https://en.wikipedia.org/wiki/Waymo*

stages. This situation where our solutions lead to new problems is a pattern and common when we are dealing with any significant degree of complexity.

In this example, it is easy to see that the unanticipated and unwanted emergent behavior listed above is not easily predictable by just knowing the *purpose* (convenience, cost-effectiveness, safety, etc.), *parts* (software, sensors, engine, body, etc.), and how the parts are *interconnected* (mechanical, electrical, signal, information, etc.) to provide the self-driving function. In fact, if we are interested in understanding the real-world behavior of Robotaxis, we would need to think broader than what is inside the Robotaxis, such as how people interact with them and think of them. We would also need to understand other things that are operating in the same environment such as emergency services, pedestrians, dogs, bicycles, traffic signs, navigation systems, fueling/charging stations, etc.

Even if we are only getting paid for designing and delivering a brand of Robotaxis like Waymo, in order to better anticipate the unwanted behaviors that might emerge, we would need to understand many things that are outside our deliverable in addition to understanding how those external things interact with each other and our deliverable. It turns out that real-world problems very quickly surpass our capabilities to gain this broader, deeper, and more effective insight. So we need *rigorous methods* (frameworks, languages, and tools), versus just some cool buzzwords, to reliably and repeatably increase our thinking and doing capabilities to a level that helps us better deal with real-world problems. We call the application of such rigorous methods to real-world problems *systems thinking*. The idea is that *systems thinking* shall help us understand the complex nature of real-world problems better so that we find the best leverage or way of steering them into a more desirable direction.

Now, why would we have enough confidence that our rigorous *systems thinking methods* would in fact work in practice? Well, from the *practice* of dealing with complex problems in a very wide range of real-world applications, it turns out that real-world complex problems have a lot in common! This commonality can be stated as a set of *Systems Principles* that are always true and always helpful when dealing with real-world problems. We will present and discuss the *Systems Principles* that are derived from practice in detail in Chapter 5. As a sneak preview here, we will use the following Systems Principles to guide our thinking about the Robotaxi example:

- *Holism Principle*: (Complex) *systems* are (much) *more* than the *sum* of their parts—*elements*, *interconnects*, and *purpose*.
- *3-Systems Principle*: *Systems* come in (at least) *three—contexts*, *systems of interest (SoI)*, and *enabling systems*.
- *Human Principle*: *Humans* impact the (wanted/unwanted) *emergence* of *all* reachable systems.

The *Holism Principle* guides us that the Robotaxi rollout problem can be more than just a combination of its known purpose and the interconnection of its elements. How much "more" depends on how complex the rollout problem is. The *3-Systems Principle* guides us to understand one or two levels outside the Robotaxi and not focus purely inside the Robotaxi components and interconnects. The *Human Principle* guides us to include humans as essential contributors to complexity, even though the Robotaxi has removed human involvement from the steering function. This textbook teaches the utilization of *Systems Principles* and methods that enforce them to help us deal with real-world problems. They are the basis for all the teachings in this book and are introduced and discussed in detail in Chapter 5.

1.15 Path to Becoming a Systems Thinker

In this book, we create a path to becoming a systems thinker. This includes learning the key systems concepts and terminology needed to understand the Systems Principles, best practices, and the rigorous methods that enable us to handle real-world problems.

Part II (Chapters 5–9) teaches the systems thinking philosophy:

- Chapter 5 lists and discusses the meaning of the *generic Systems Principles.* Applying these principles improves our chances of understanding and interfering with real-world problems.
- Chapter 6 uses cartoons, card games, sports, and movie clips to help us develop an eye for the principles at play around us.
- Chapter 7 shows how to apply these principles and use them to guide our thoughts, actions, and behaviors.

- Chapter 8 discusses the issue of shortage of critical systems thinking talent to help individuals and organizations follow a more fruitful path to addressing this challenge that is also an opportunity.
- Chapter 9 lays out pathways for creating a culture that fosters systems thinking.

Systems thinking is about *doing* as much as it is about *mindset* for people of all technical and nontechnical backgrounds. In other words, this book is not just about learning the systems thinking terminology, philosophy, or buzzwords. It is also about using rigorous methods that drive the correct actions and behaviors at the individual and organizational levels. So the talk is balanced with the walk! Part III shows how rigorous methods (such as model-based systems thinking) enable us to apply the *Systems Principles* to real-world problems in various domains. Now, before in-depth training on the systems thinking philosophy can begin, we need to better understand why it is essential when dealing with real-world problems. The rest of the chapters in Part I use easy-to-follow and often well-known real-world examples showing why systems thinking is essential and what happens when it is lacking.

1.16 Reflection on Chapter 1 Material

Systems thinking can be used to navigate the complexities of real-world problems. It offers a pragmatic and hopeful way forward for understanding difficult-to-understand (and predict) issues from personal challenges to global challenges. We have a societal need for adept systems thinkers, presenting an opportunity for us to be a catalyst for change. Becoming a skilled systems thinker will change how we see the world around us and increase our odds of success when facing complex challenges.

We began this chapter by introducing some basic systems concepts and terminology necessary to understand the systems thinking philosophy. We then used Tom/Jerry cartoons and the Waymo story to show that quick solutions resulting from linear or simple cause/effect thinking are not the best, particularly when the challenge is complex or when there is much more to the challenge than meets the eye.

The rest of the material in Part I of this book gives us the background we need so we are better prepared for Part II, which teaches the systems thinking philosophy. Let's go!

References

1. Adcock, R., et al. (2024). Principles of systems thinking. In *Guide to the systems engineering body of knowledge (SEBoK)*. https://sebokwiki.org/wiki/Principles_of_Systems_Thinking.
2. Forrester, J. (1992). *From the ranch to system dynamics: An autobiography*. A. G. Bedeian (Ed.) (Vol. 1). Management Laureates. JAI Press.

3. Gharajedaghi, J. (2006). *Systems thinking: Managing chaos and complexity: A platform for designing business architecture*. Elsevier. ISBN: 0-7506-7163-7.
4. Harvard School of Public Health. *Systems Thinking*. n.d. https://ahpsr.who.int/what-we-do/thematic-areas-of-focus/systems-thinking.
5. Senge, P. (2006). *The fifth discipline: The art and practice of the learning organization*. Random House Books.
6. Senge, P. M. (1994). *The fifth discipline fieldbook: Strategies and tools for building a learning organization*. Crown Currency. ISBN: 978-0385472562.
7. Sterman, J. D. (2000). *Business dynamics: Systems thinking and modeling for a complex world*. Irwin McGraw-Hill.
8. Stroh, D. P. (2000). Leveraging change: The power of systems thinking in action. *Reflections, 2*(2), 51–66.
9. Stroh, D. P. (2015). Systems thinking for social change: Making an explicit choice. *Reflections, 14*(3), 35–42.

Chapter 2
Why Is Systems Thinking Essential?

Abstract Systems thinking requires an intentional mental practice that looks at multiple perspectives to understand real-world complex problems. It encourages individuals and organizations to address complex problems in a holistic and comprehensive manner. It involves understanding that various components of a system are interconnected and that changes in one area can ripple through others, often in unpredictable ways. By adopting a systems thinking mindset, we can better anticipate the potential consequences of decisions (or actions), identify leverage points for change, and develop solutions that consider the broader context rather than just isolated parts. Systems thinking attempts to cause a shift from a linear, cause-and-effect perspective to a more dynamic understanding of how things work, including feedback and circular effects. Systems thinking is not just about thinking and using fashionable terminology to do the same. It encourages a balance between thinking and doing. Employers across disciplines and industries increasingly recognize systems thinking as a set of required skills. It can be an essential part of a strategy for us to aspire to drive positive and sustainable change. Systems thinking positions us to unlock opportunities that were missed by classical, linear, and compartmentalized methods. Applying systems thinking in projects expands your resume of work experiences, opens directions for significant contributions within your teams and organizations, and will enrich your professional portfolio.

2.1 Inadvertent Impact of Culture and Education

Most students have been taught by school systems to follow traditional thought processes to analyze and solve problems where a problem is broken up into manageable chunks; each piece is analyzed, better understood, and improved in the hope that the whole system is improved when these are added back together. This linear compartmentalized problem-solving method was tried and tested in the past and works for simpler problems. As society and technology have progressed, so has the complexity of systems and, therefore, problems. Systems are increasingly inherently interconnected within elements and with the environment.

K. Eftekhari Shahroudi et al., *Practical Systems Thinking*,
https://doi.org/10.1007/978-3-031-89394-0_2

In navigating the intricate landscape of our world, we encounter a web of interconnected problems. Understanding the interconnected nature of different system elements, or various systems coexisting and interacting, necessitates an intentional mental practice that looks at multiple perspectives to solve these problems. This intentional and holistic approach is to acquire a set of skills and perspectives that are invaluable in navigating the complexities of life, both personally and professionally.

Unfortunately, the inadvertent impact of our cultural, secondary, and higher-level educational systems is that many of us, unfortunately, develop misconceptions, mental models, thinking habits, and methods that are not suitable for facing complex real-world challenges. These are identified and explained in detail in Section 5.1. Sections 5.2 to 5.10 teach a set of principles that form the basis for systems-oriented mental models that are more effective for understanding and creating solutions for complex real-world problems.

2.2 What Does It Mean to Be a *Systems Thinker*?

To be a systems thinker is to use a set of *Systems Principles* to guide our thinking, actions, decisions, and use of methods to enhance the ability to understand and possibly improve real-world complex problems. A systems thinker is better equipped to address challenges that span multiple disciplines, domains, roles, organizations, and diverse facets of life. It is a conscious choice: a way of thinking, making decisions, and doing that everyone does not already have, but everyone can learn.

Becoming a systems thinker will transform the way we approach difficult real-world problems or challenges. It requires the investment of time, effort, and practice on complex real-world problems, but once we have the principles and methods under our belts, it is just like riding a bicycle!

Chapter 7 gives an in-depth discussion of the *ways of the systems thinker* that shows how our thinking and acting can be aligned with the Systems Principles.

2.3 Creating Better Solutions

Imagine a Robotaxi *behavior* that is always perfectly (i.e., 100%) predictable and always perfectly (i.e., 100%) aligned with the stated purpose in *all* possible situations and *all* possible operational scenarios in *all* cities from the moment its concept was generated to prototype, to service introduction, to retirement, and to recycling at the end of service life! What kind of thinking, mental discipline, and approach would we have to follow to get closer to this level of perfection?

Obviously, an oversimplistic mental model of the Robotaxi that ignores all the following complex issues is not the way:

- Ignoring that the structure, behavior, and purpose of the Robotaxi influence each other. For example, how the Robotaxi behaves to avoid collisions can lead to increased congestion, reduced efficiency, and overall safety that is against the intended purpose of the Robotaxi.
- Ignoring that there are other things, actors, and systems outside of the Robotaxi that can adversely impact the Robotaxi or be adversely impacted by it.
- Ignoring the emergence that results from the interaction of the Robotaxi with other actors in the same environment, particularly the humans figuring out new ways of utilizing the Robotaxi like drug use.
- Ignoring that the cause of issues impacting the Robotaxi may not be at the same time and location as when problems are noticeable. For example, activist perception of the city prioritizing politics and money above safety did not occur immediately after rollout but took time to brew until they started to shut down the Robotaxi by simply putting traffic cones on their hood.

Now, obviously, testing solutions in the real world (and not in a controlled lab) is the ultimate way to find problems. However, fixing issues after service introduction is orders of magnitude more expensive than early on when the design and manufacturing are not yet baked in. When we look at the unwanted issues related to the Robotaxi rollout in San Francisco in Fig. 1.8, it is not a stretch to think that balancing our efforts and focus inside and outside of the Robotaxi would have helped us predict at least some of the unwanted emergent behavior earlier and solved them already before service introduction where fixes are very expensive. For example, we could have predicted that there might be an activist group emerging, particularly since Cruise and Waymo ultimately replace paying jobs that impact the livelihoods of people in the short term. It may have been harder to predict the human response of putting traffic cones on the hood or wearing T-shirts with a big red stop sign, but even that could have probably been predicted by a group of people that have a hacking mindset to try and find all the ways the proper functioning of the Robotaxi can be impacted. Note that we are not claiming that "all" complex emergence is knowable or predictable, but the real-world story of the Robotaxi shows that a good portion often can be anticipated, predicted, and possibly mitigated earlier when fixes are cheaper.

2.4 Understanding Real-World Challenges and Problems

There are many real-world examples all around us where a better way of thinking to balance our efforts and focus would have helped us anticipate unwanted consequences long before they actually became known. LEDs are much more efficient than classical lamps that were previously used in car headlights. They are very bright relative to the energy they consume and enable a lot of design flexibility to enable more advanced functions at a relatively lower cost than classical lamps like halogen projectors. In particular, LEDs are guaranteed to work in cold temperatures down to

Fig. 2.1 Several manufacturers did not predict the issue of LED headlamps icing up in cold climates. Systems thinking would have easily helped

20 F. Classical lamps (like the halogen lamp), in comparison, work well when they reach their ideal operating temperature, but their function is limited until they warm up.

Given all the benefits of the LED, it seems like a no-brainer that changing headlamps to LEDs would be a great improvement. Yet, when they were first introduced, many users in cold climates experienced a new issue that they never had before: headlamps would ice up as shown in Fig. 2.1. It turns out that since the LED is very efficient, the heat dissipated from them is also much less than the classical lamps used previously. The rejected heat from the old halogen projectors was enough to overcome icing in cold weather conditions. But switching to the LED that by itself has better low-temperature performance, when integrated into the headlamp assembly, has problems in cold weather! This was a real-world issue that was not anticipated or predicted before its introduction into service by many manufacturers who have been building millions of cars working in severe cold climates such as Colorado, Wyoming, and Alaska for decades.

Like the Robotaxi, it is not a stretch to imagine that a few smart and experienced individuals who have designed and built successful cars previously and have access to known specifications for LEDs and halogen projectors would have been able to predict the icing problem. Yet, in reality, they did not. What is it that stops us from anticipating or predicting seemingly simple issues like this?

One cause is that our culture and education have inadvertently given us a mental model for solving problems: the dominant way to solve problems is to divide them up into smaller parts and allocate those parts to different people or groups who then focus their efforts on those parts. If the parts are too large, say, for individuals or groups to handle, then keep dividing them into smaller parts until they can be handled. The idea is that the whole car would improve by improving the individual parts without worrying too much about how the parts interconnect, interrelate, or interact to create a behavior that is not a simple combination of the way the individual components behave. Pure internal focus on the parts is also doomed to failure because often the interactions that cause the non-simple emergent behavior come from the interaction of the parts with things that are outside of the system of interest, not inside. In this example, the unwanted icing behavior is from the way the LED lamp and the headlamp lens interact with moisture, snow, and ice in cold weather conditions that are outside of the car. Now, it would not be in the best interest of car manufacturers to claim that they are not responsible since the cold weather, ice, and snow are located outside of the car and not directly generated or controlled by the internal parts of the car. Why? The interaction of this position with customers is likely to lead to another complex emergent behavior where customers start migrating to competing products that do not have this view.

Unfortunately, an overly simplified mental model of the best way to approach real-world problems has become second nature or the main mode of operation: without realizing it, many imagine their problem to be a simple linear problem for which they expect that the divide-and-conquer approach with internal focus on the parts will always work. Well, here is a real-world situation where the LED part is superior to the halogen lamp in many important ways. Yet, after this update, many people in cold climates experienced problems they never had before!

Real-world problems are not required to fall perfectly and neatly into the way we like to divide the problems. For example, it would be nice if the icing problem was predicted and solved by the LED lamp expert. But the problem wasn't predicted by LED experts and also not solved by the LED experts because, from their point of view, there is nothing wrong with the LED component itself! It is interesting that multiple companies and car models experienced icing after the LED update, so we can't blame the problem on a particular individual or company. It is clear that no matter how we further improve the efficiency, cost, life, and performance of the LED part alone, we cannot solve this seemingly simple real-world problem, because the problem occurs because of the interaction of the LED with the headlamp cover: i.e., less heat dissipated by the LED is not enough to melt the ice and snow buildup on the headlamp cover.

2.5 Complexity Is in the Eye of the Beholder!

Let's call real-world problems or solutions to real-world problems that are *difficult* to understand, predict, know, anticipate, or solve *"complex."* All humans already deal with complex problems as babies. Learning to walk, talk, and run are among our first complex challenges to overcome, but to someone who is already walking, talking, and running very effectively, these are no longer complex or not a big deal. So can we answer this simple question? Is walking, talking, and running a complex problem? The answer obviously depends on many factors: whose perspective, which environment, what enabling tools we can use, what is the purpose of walking,

Fig. 2.2 The A380 has a stellar safety and reliability record. Is the A380 complex (hard-to-predict emergent behavior) or just complicated (many parts and interconnects)? The answer is it depends

what are the constraints on acceptable language, etc. Walking in a shopping mall is different than walking on a steep, jagged slope on Mount Everest. All we have to do is to change the environment, enabling tools, players, purpose, and perspective for a real-world challenge that is simple to become complex, and vice versa.

In the LED example above, it was in the car manufacturers' interests to anticipate and mitigate issues related to safety. Several car models did not effectively anticipate or predict the icing problem due to the lighting technology change to LEDs. So from their perspective, the icing problem was a complex problem, but from the perspective of a systems thinker who is not purely focused on the parts but also focused on how the parts interact with each other and the environment, this is a relatively easy problem to predict and fix earlier prior to introduction into service.

Now, consider the double-decker Airbus A380 shown in Fig. 2.2, one of the largest commercial passenger transport airplanes still flying. Is the A380 a complex system? Well, the finished product that is in service today has a stellar safety track record at the time of writing this text. It reliably and safely transports many passengers around the world without serious or major issues. Some airline operators used it as their flagship. The only major issue was that the A380 was so large that it was not always possible to fill all the seats with passengers on all routes, losing potential revenue. So the A380, which is one of the largest passenger aircraft ever built, has enjoyed a very predictable behavior and performance after it entered into service. Since it operates very predictably, like a finely tuned Swiss watch, then some would argue that the A380 is just complicated (many parts, interconnects, operating scenarios, etc.) and not complex. However, the A380, like all other systems, has a life cycle. When it was merely a concept, there were a lot of real-world effects that were very hard to understand and predict.

The A380 took about 5 years and several thousand employees from the creation of the A380 concept to its first flight. Even at first flight, it was not totally flawless, and it still took a lot of testing, analysis, and development to solve all the challenging real-world problems associated with its design and setting up the manufacturing line, certification, pilot training, etc. So developing the A380 was a very complex challenge with many real-world effects that were not predicted initially. So is the A380 complex or just complicated? The answer is that it depends on the purpose, perspective, environment, operational scenarios, area of responsibility, human influence, dominant mental models, interactions inside and outside of the A380, etc.

We saw earlier that for a linear compartmentalized thinker, the problem of switching to the LED becomes a real-world complex problem. In other words, the mindset does not enable predicting the issues that emerge from the interaction of the parts. But the same problem appears simpler when we balance our efforts and focus on the purpose (that includes operating in cold environments), the parts, "and" the interaction of the parts. So our mental models and how we define and approach the problem determine how complex or simple our real-world challenges become. Two people can be talking about the same challenge or system, and they can both be right if one believes it is a complex problem and the other believes it is a complicated or simple problem. The simple statement below helps remind us that complexity is not an absolute characteristic of a problem or system:

Complexity is in the eye of the beholder!

2.6 Emergence and Sources of Complexity

Systems thinking texts often use a phrase similar to the following: *A system is* more *than the sum of its parts*. If a system was equal to the sum of the parts, then the properties and the behavior of the system are a simple linear sum of the properties of the parts. For example, the properties of 1 kg of rice are very simply and linearly related to the properties of 100 g (0.1 kg) of rice. The former is just bigger, heavier, and occupies a volume that is more or less linearly predictable from the volume of the latter. The number of rice kernels in 1 kg is approximately ten times the number of rice kernels in 100 g. The taste and texture of 200 g of cooked rice is predictably the same as the taste and texture of 100 g of cooked rice, assuming we follow the same recipe for both and proportionally adjust for all the other ingredients such as water, salt, oil, etc. So for the purpose of taste, 200 g of rice is *not more* than 100 g of rice following the same recipe.

Now, consider the survivability of, say, one zebra versus a large herd of zebras in the predator/prey context. The zebra skin markings are such that they merge together in a herd, making it very hard for the lions to pick out a specific zebra to target (see Fig. 2.3). Furthermore, a lion targeting a member of a large running herd faces the added risk of getting bit, hit, or trampled by lethal hoofs. So in the context of survivability, the tight herd is a lot *more* than just a larger number of zebras. Linear or oversimplistic thinking might suggest that a large number of zebras would actually increase the lion's chances since the prey is more plentiful. There is no *safety in numbers* if the zebras spread out and do their own thing individually. So the effect of zebras collaborating in a coordinated herd of zebras is the opposite of what linear or oversimplistic thinking might conclude!

In other words, the emergent behavior and survivability of a herd of zebras are not simply or linearly related to those of a single zebra. If we call the herd a system and the individual zebras the parts that collaborate together towards a common purpose of surviving, then the complex emergent behavior is the result of the interaction of the parts, which is a drastic improvement in the survival rate of zebras.

Fig. 2.3 Merging stripes in a herd of zebras make individual zebras hard to target by lions

Fig. 2.4 YouTube story of fighting mosquitoes with DDT leading to collapsed roofs and rodent explosion in Borneo

Figure 2.4 is a YouTube story of trying to control the outbreak of Malaria, where using DDT on a grand scale to kill mosquitoes leads to collapsing thatched roofs and much more! The collapsing roofs were because the DDT also killed the bees that controlled the population of larvae that damaged thatched roofs. This story is discussed in more details in Chapter 5 and shows yet another source of complexity that is not emerging from the interaction of parts but from the interaction and interrelationship of different (eco)systems (i.e., feedback loops between systems) coexisting in the same environment.

The Robotaxi example of Fig. 1.8 showed yet another type of complexity where the interaction with humans inside and outside the Robotaxi leads to the emergence of complex behavior that was hard to predict early prior to its rollout in San Francisco. It is interesting that humans can be the most significant influence or element involved in creating wanted and unwanted emergent complex behavior even in the case of autonomous systems that intend to cut them out!

2.7 Perspective and Way of Thinking Matter

Figure 2.5 shows two very different ways to view a tree. The left is a microscopic type of view where we divide the tree into smaller parts to study them deeper. The right view shows the tree inside an ecosystem where we study how the tree integrates with and interacts with other things outside the tree: animals, soil, climate, etc. The question is which is a more useful and practical view of the tree? Which do we prioritize: looking inward or looking outward? The answer is that they are both valuable! It depends on the purpose of the study. However, the generic correct answer is that we should balance our efforts and focus among several most appropriate perspectives.

Unfortunately, some texts appear to confuse what systems thinking suggests here: They believe systems thinking is only about the integrative view like the one on the right side of Fig. 2.5. But as we saw in Section 2.6, complex emergent behavior can come from the interaction of internal parts (the left view), or it can come from the interaction of the system with other systems in the environment (the right view), or it can come from human influence or several other means. So systems thinking cannot possibly prioritize one of these views over the other. It simply encourages us to not just focus on one standard perspective but actively seek several most useful perspectives to balance our efforts and focus.

Now, watch the first 1 minute and 12 seconds of the Amazing Card Trick YouTube video in Fig. 2.6. The goal is to figure out how the cards changed color, and for most people not into card tricks, it is hard to figure it out even if we just focus on the cards. Did the selective focus on the cards stop you from seeing any other changes? It turns out that there were four changes in the first 1 minute and 12 seconds that most people miss on first viewing. Now, watch the rest of the video that uses a broader, less deceptive perspective. It all becomes easier to understand with the new better perspective recorded on a different camera. This video shows that we should actively seek better perspectives and balance our efforts among them because perspective impacts how difficult or complex the problem appears.

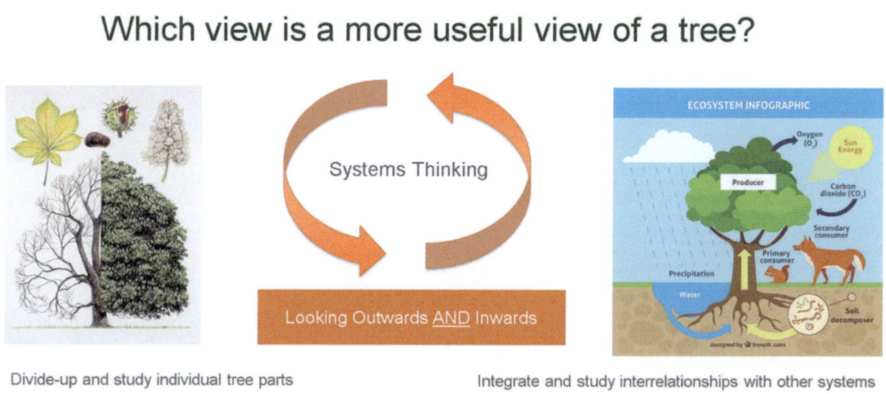

Fig. 2.5 Systems thinking does not prioritize the best view: it simply requires that we balance our efforts and focus among several useful perspectives

Fig. 2.6 Perspective matters: the perspective that the cameraman of this YouTube video imposes in the first minute and 12 seconds intentionally makes it harder for the viewer to understand how the cards changed color, in addition to missing three other changes. The rest of the video uses a different perspective that makes it very easy to understand the trick

2.8 The “s” in System*s* Thinking

Did you notice the “s” in system*s* thinking? Why do we not just use the singular form system thinking without the “s”? Well, complex emergent behavior can often come from the interaction of the system we are interested in with other systems that coexist in the same environment. The Borneo Malaria YouTube video was an example of this where using DDT to kill mosquitoes had far-reaching consequences on several other species and thatched roofs that are not directly related to the mosquitoes that were the target for the DDT decision. The Alexander the Great story of Chapter 5 is another great example where Alexander managed to defeat the far superior Persian Navy not by facing them head-on, but by attacking the island of Tyre, which was a supply base for the Persian Navy.

So we have multiple systems, multiple perspectives to view those systems, and interactions within and among multiple systems, hence, the “s” in system*s* thinking.

2.9 Is Systems Thinking a Waste of Effort for Simple Problems?

Some may argue that systems thinking may be a waste of effort or overkill for problems that are simple, linear, and routine. If there was some sort of simplicity guarantee for real-world problems, then we would probably agree with this statement. Unfortunately, real-world challenges in most nontrivial cases do not come with such guarantees, and typically, we see the opposite problem of underestimating the potential complexity of the challenge. An example is the 737

MAX problem where after certification and entry into service, a miscalibrated angle of attack sensor worth only a couple thousand dollars caused two fatal crashes, an estimated 50 billion dollar damage, and the knock-on effect to other non-737 MAX projects at Boeing, airline customers, and the air transportation system as a whole. If best-in-class companies with decades of stellar experience and pedigree can severely underestimate the complexity of their real-world challenges, then this can happen to anybody or any organization.

Several 737 models, in addition to other models like the 747, were the workhorse of the air transportation industry and had a stellar track record before the 737 MAX. They were all very predictable and reliable and transported millions safely to their destinations. Some may argue that since the architecture and behavior of the 737 were very knowable and predictable, then the 737 is just complicated (as it has many parts and interconnections) versus complex, and therefore, systems thinking may not be required. The 737 MAX example is described in more details in Chapter 4. There, we will see that upgrading to new engines that were very knowable and predictable also caused minor changes to the flight controls. This change interplayed together with decisions from various technical and business departments, including the FAA, to create a major industry and airline problem. So, a problem that appeared to be knowable, predictable, and business as usual to all the different departments, experts, and organizations suddenly exploded into a fully complex major issue, causing a lot of damage. All it took was a few minor changes that in themselves were small and appeared benign locally, but together caused a huge complex emergence.

Our real-world challenges and the solutions created to deal with them are becoming increasingly complex. Societal challenges like homelessness are increasing despite increasing expenditures and efforts to control them. Technical systems (aircraft, cars, phones, etc.) are increasing in complexity because of new technology trends and an explosion in the number of CPUs, sensors, actuators, software, and hardware modules, where the number and strength of interactions between the parts are increasing. Sociotechnical systems like social media platforms are also increasing in complexity not just because of the growing numbers of users but also the strength of coupling between the groups of users and their impact on commerce, politics, world affairs, population mental and physical health, etc.

So unless we are dealing with super clear routine trivial problems that neatly fit into a narrow discipline, then chances are that in most real-world scenarios, we are better off adopting the systems thinking mindset. We cannot take simple routine problems for granted because all it takes is seemingly minor changes that ripple through the system and amply each other to cause a huge potentially unwanted emergent problem.

2.10 Reflection on Chapter 2 Material

Systems thinking requires an intentional mental practice that looks at multiple perspectives to understand real-world complex problems. Problems must be tackled holistically and comprehensively. This involves understanding that various system components are interconnected and that changes in one area can ripple through others, often in unpredictable ways. By adopting a systems thinking mindset, we can better anticipate the potential consequences of decisions (or actions), identify leverage points for change, and develop solutions that consider the broader context rather than just isolated parts. Systems thinking attempts to cause a shift from a linear, cause-and-effect perspective to a more dynamic understanding of how things work, including feedback and circular effects. Systems thinking is not just about thinking and using fashionable terminology to do the same. It encourages a balance between thinking and doing. This balance is an important point that we emphasize in many places in this book. Systems thinking positions us to unlock opportunities missed by classical, linear, and compartmentalized methods. Applying systems thinking in projects expands our resume of work experiences, opens directions for significant contributions within our teams and organizations, and will enrich our professional portfolio. In this chapter, we discuss why systems thinking is essential. Next, let's dive into real-world examples where a lack of systems thinking caused more issues.

Chapter 3
Real-World Examples Showing Lack of ST

Abstract Here, we discuss three real-world problems that would have benefitted from a greater application of systems thinking. The first example is about a social worker tasked with a decision on whether to remove an obese child from his family for health reasons. The second example is the homelessness problem that has exploded in major cities in the USA despite increasing efforts and roughly $3 billion in recurring expenditures to stabilize it. The third example is the Boeing 737 MAX problem, where a miscalibrated angle of attack sensor, a very minor and well-understood element of flight controls, interacted with other well-understood technical and administrative elements to cost lives and more than 50 billion dollars in damage to the manufacturer. All are examples of emergent hard-to-predict behaviors that result from the interplay of several causal relationships.

Real-world complex problems consume significant amounts of effort and resources that often do not work due to insufficient levels of systems thinking or a limited mental model of the problem. Our intent is not to blame the particular organizations involved. The examples we picked are not the worst real-world examples and they are also not unique. We realize that limited mental models leading to bad decisions and performance can occur anywhere and can impact everyone: even best-in-class organizations or individuals with impeccable track records are not immune to implementing quick solutions that make matters worse over time. The types of problems that we point to in this chapter impact many industries and organizations. Systems thinking is not a guarantee against bad decisions, but it improves our mental models to help anticipate the unwanted real-world impact of our decisions.

3.1 Childhood Obesity

The causes of childhood obesity are straightforward, right? Parents are not promoting a healthy diet, and the child has ample access to junk food. The child has no concept of the importance of a healthy diet. Parents have no energy to fight the

K. Eftekhari Shahroudi et al., *Practical Systems Thinking*,
https://doi.org/10.1007/978-3-031-89394-0_3

food battles and serve "kid meals" of chicken nuggets and macaroni and cheese. Families do not commit to having family dinners together. The child is lazy and plays too many video games. Physical activity is not a priority. As soon as a child does not want to participate in a sport anymore, the parent no longer forces the issue. The child uses food as a source of comfort. The child is genetically predisposed to obesity—it just cannot be helped; it is the way they were built.

Of course, this is by no means a comprehensive list of causes. Childhood obesity is inherently a complicated problem that is forecasted to grow as seen in Fig. 3.1 and cannot be reduce to a single cause for each case. To level set, the CDC defines obesity as a body mass index (BMI) at or above the 95th percentile BMI-for-age growth charts [1]. With linear thinking, each cause is addressed independently, and solutions for that cause are developed. To encourage a healthy diet, schools are serving healthy choices in the lunchroom, limiting the availability and marketing of unhealthful foods and sugary drinks, and making water available to students throughout the day [6]. To promote physical activity, elementary schools have required physical education (PE) programs, and after-school community organizations offer various sports programs. The list goes on and on for the various school and community programs that are available to kids and families. However, childhood obesity has been on the rise, as shown in Fig. 3.1, for several years now despite the best efforts of schools and community organizations, so linear thinking would lead us to believe that it must be the fault of parents. In 2011, a third-grade

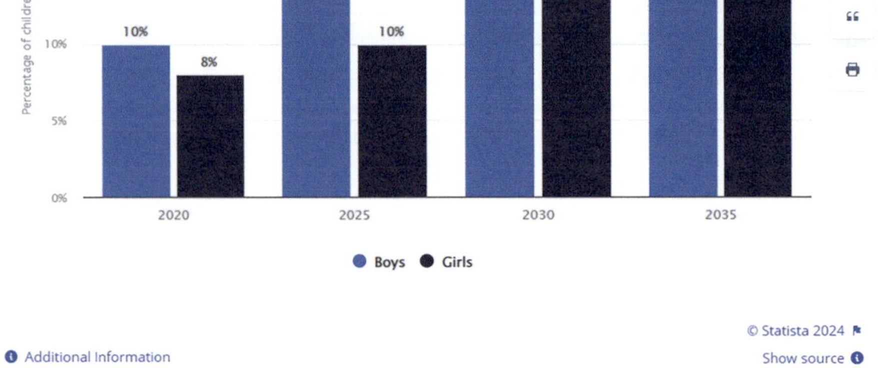

Fig. 3.1 Statista: percentage of children and adolescents aged 5 to 19 years worldwide who were obese in 2020 and forecasts to 2035, by gender

student who weighed more than 200 pounds was removed from his home and placed in foster care because officials concluded that the mother was to blame [3].

The following story was shared by Di Lacey, DHSc, Vice Provost of Strategy, Growth, and Transformation at Western University of Health Sciences (personal communication, August 13, 2024).

In 2013, Charlie was a morbidly obese third grader living in Portland, Oregon. His primary care physician had repeatedly counseled Charlie and his mother on the severity of his health issues and the immediate need for improved exercise and diet. Every year, his mother assured the physician that she would implement the necessary changes. But as Charlie's weight continued to increase and his health status decreased, the physician was forced to assume medical neglect on the part of his mother and consider the recommendation of placing Charlie in foster care. A wise social worker Jennifer decided that it would be prudent to make a home visit before executing this extreme call. When Jennifer arrived at Charlie's house, she noticed a park within viewing distance of the home. She immediately queued up for the lecture she would give Charlie's mom. It was obvious to her that the mom needed to insist that Charlie play in the park instead of sitting around the house watching TV. When Jennifer got into the house and had a minute to sit with Charlie's mom, she immediately launched into her spiel about how Charlie needed to spend more time playing at the park and getting exercise. Charlie's mom responded by saying that if Charlie was going to die, then he was going to die at home with her. Confused, Jennifer asked why he was at risk of dying at the park. Charlie's mom said, "Look again, out this window, at the park. There is a large tree that blocks me from having a direct view of the park so I cannot see Charlie out there. Neighbors let known aggressive pit bulls freely roam in the park. The playground is littered with needles, and everyone around here knows that drug dealers frequent the area. It is less safe for Charlie to play at the park than to stay here with me."

Jennifer realized the immediate answer was not to remove Charlie from his home. She also realized that problem was more complicated than she initially thought and that to make the park safe for Charlie and all the other neighborhood kids to play at, multiple agencies needed to work together—local law enforcement to mitigate drug trafficking, animal control to enforce leash laws, the parks and recreation department to ensure the safety of public park infrastructures, city management to maintain low foliage on sidewalks and medians to provide line of sight for parents and guardians, etc. This was a lesson in the importance of abandoning a linear thinking mindset for a more helpful systems thinking mindset of zooming in and out, being able to look at both the big picture and the details when identifying potential solutions [10]. Systems thinkers seamlessly toggle between looking at the whole forest and each tree in the forest. Getting stuck in "zooming in" mode is the same as linear thinking.

Childhood obesity is a complicated problem that linear thinking cannot solve. Each case is like Charlie's; there are multiple contributing factors and feedback loops that need to be considered and understood before rash decisions are made or ineffective solutions are implemented.

Granted, individual organizations are doing their best within their sphere of influence, but they will never be able to fully address the problem until they abandon their linear thinking ways and adopt a systems thinking mentality.

3.2 The Homelessness Problem

Unlike simplified problems used to teach and evaluate us in school, complex real-world problems do not fit neatly inside narrow disciplines (like math, physics, chemistry, etc.). They often transcend classical narrow disciplines. They are also often *ill-defined* or *ill-posed*: it is rare that we get precisely the right amount of reliable information that we need to understand or solve a complex problem. Sometimes, there is too much information, while at other times, there is too little or unreliable information.

3.2.1 Background

In many situations, the information may be counterintuitive or incorrect. For example, Fig. 3.2 shows a sharp drop in homeless in 2021, which is not very believable because none of the many factors that might impact homelessness (jobs, prices,

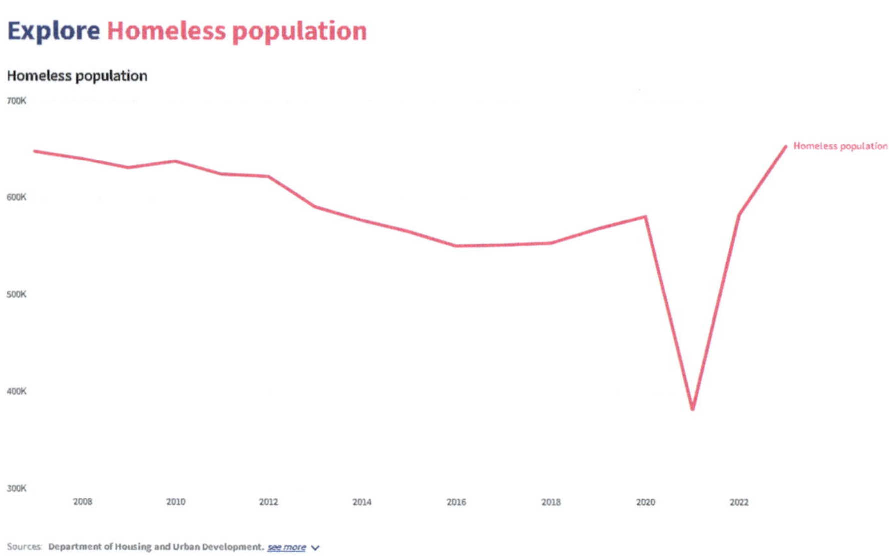

Fig. 3.2 How *real* is this pattern of behavior in US homeless data? Was the drastic improvement in 2021 real?

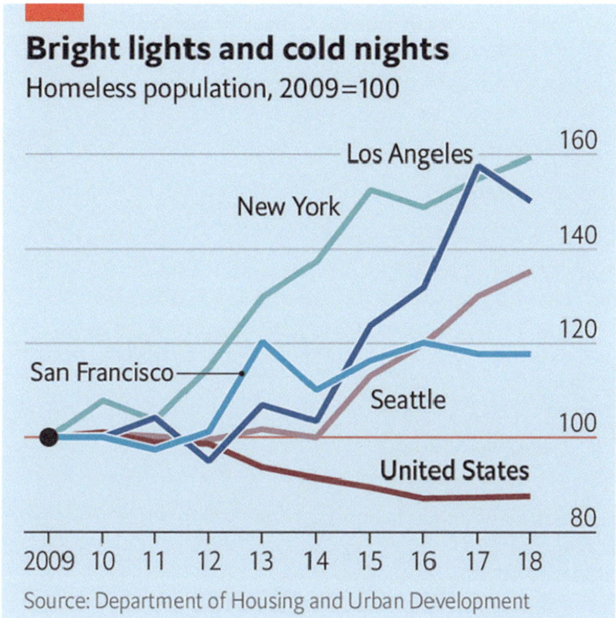

Fig. 3.3 Can we conclude homelessness is improving or stable overall and only a few cities have issues?

health, addiction, shelters, available homes, etc.) moved in a favorable direction in 2021! Due to COVID and shutdowns, many locations that were collecting and populating homelessness data skipped a year, so the drastic 2021 effect that the data shows in 2021 is not authentic or reliable!

Figure 3.3 from *The Economist* article suggests that homelessness is improving overall, but (some) cities are getting worse. For example, based on this article, one may conclude that the efforts and expenditures to combat homelessness in the USA are generally working (i.e., stabilizing homelessness), except in major cities that have some special additional issues and factors. But this would have been the wrong conclusion; as we now know, in 2024, the overall US homeless numbers have grown every year from 2018 until the time of writing this book. This shows that when dealing with a complex problem, we can very easily arrive at a wrong mental model or conclusion by directly projecting or extrapolating from the available data and prevailing opinions.

3.2.2 Predicting a Future by Eyeballing Data

Instead of projecting a baseless conclusion directly from data, systems thinking encourages us to look for structural causes of previous and potential future emergent

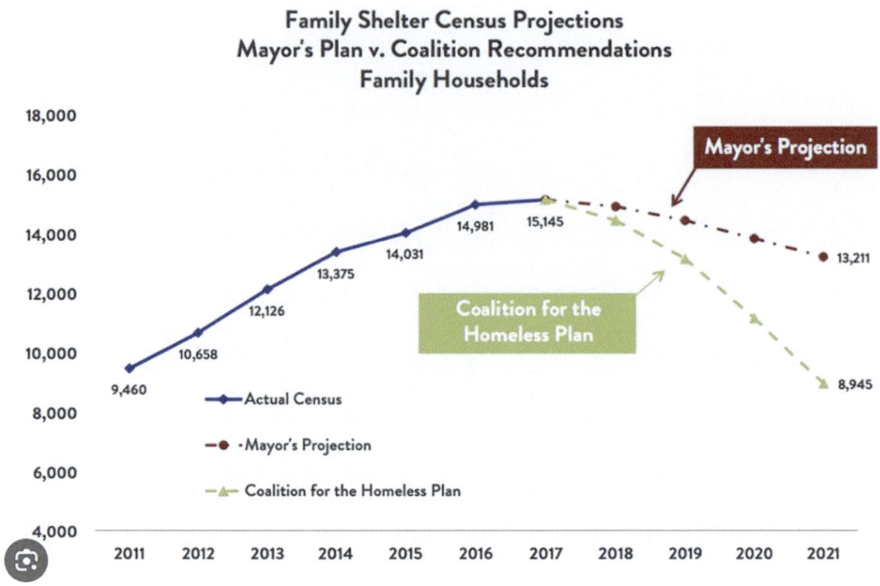

Fig. 3.4 Can we conclude homelessness is improving or stable overall and only a few cities have issues?

behaviors. These structural causes come from identifying all the relevant parts of the systems that play together in an environment and how they influence each other. This approach improves our chances because, with systems thinking, we are able to better predict the precursors or conditions that give rise to unwanted emerging behaviors.

In complex real-world situations, it is often the case that short-term gains are not sustained or become worse over time. So rather than thinking that only a few cities have special issues, as shown in Fig. 3.3, versus a stable and controlled overall US homelessness, systems thinking raises the possibility that the behaviors may be interrelated: For example, it is possible that these cities are more sensitive and show growth much sooner than the rest of the country but impact the overall number after some delay: i.e., homelessness at cities are not disjointed from the overall US homelessness, but they may be influencing each other over time. We should put our efforts and resources into discovering the influence of various behaviors on each other versus short-term solutions based on thinking we can directly predict the future by eyeballing available data!

Figure 3.4 shows another interesting effect for New York City, where homelessness is a large and growing problem: it appears that in 2017, both the Mayor's office and the Coalition for Homelessness were projecting future reductions in homeless numbers, only differing by how big the projected reduction was. However, Fig. 3.5 shows that they were both wrong and drastic increases in homelessness actually happened post-2017 with the exception of 2021, where data is unreliable.

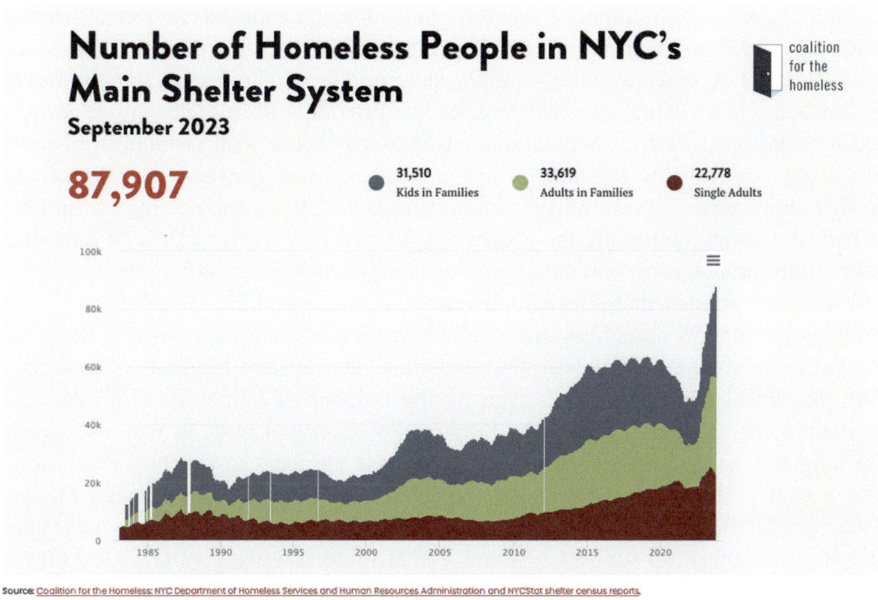

Fig. 3.5 Can we conclude homelessness is improving or stable overall except a few cities with special issues?

So homelessness is similar to other real-world problems where the complexity makes both the problem and the solutions hard to understand and predict. Expecting that a reduction in the overall US homeless number would continue into the future past 2017 or that it would impact NYC in a favorable direction turned out to be very wrong, despite the best intentions of people responsible for understanding and fixing the problem.

3.2.3 Quick Solutions

Another issue with quick conclusions and opinions: If we look deeper, we realize that the statistics used in the above figures are related to but are not really measuring homelessness directly. They count the number of people who interact with shelters or local government offices to receive some services on a particular day in the year! So, for example, they count the utilization of shelter facilities on January 1. This excludes those who are not, for example, taking advantage of shelters or live in areas, where there are not a lot of government offices to interact with. It also excludes weekly, monthly, and seasonal patterns! So it is very possible that the reduction and stabilization effect of the overall US homelessness shown in Fig. 3.3 is purely an artifact of how this is measured and not really reflecting the overall US homelessness.

There have been a number of previous well-meaning solutions that inadvertently made the problem worse. For example, compensating service providers based on the occupation of beds in shelters simply increased the shelter usage and number of shelter beds. Is focusing on rewarding service providers for increasing the number and occupation of shelter beds aligned with our purpose to find long-term, more sustainable solutions to the homelessness problem? The answer is no! Pure focus on this short-term solution will look good in terms of just measuring the number of beds but is not addressing the causes that impact the incoming flow of homeless people and the outgoing flow into sustained long-term housing solutions.

There are articles that suggest, for example, that spending $30 billion a year would solve the US homelessness problem, suggesting that the money is there but the willingness is not! The way they calculate the amount required is to simply multiply the housing cost per person by the number of homeless. However, this is likely to be another quick linear thinking solution that may, in fact backfire in the long run and potentially cause the homeless numbers to explode even more. The reason is that homelessness is a complex interplay of many societal factors: poverty, addiction, family situation, costs, physical and mental health issues, jobs, climate, well-meaning support systems, etc. If tomorrow we give US $100,000 to every homeless person in Fort Collins, a small portion of them would probably climb out of homelessness in the long term, but it would not solve the issue for the majority in the long term because it does not address many of the factors and systems that interplay, giving rise to the complex behavior of homelessness over time. Furthermore, as news of this policy travels, many homeless from other cities would move to Fort Collins to benefit from the $100,000 benefit.

3.3 The 737 MAX Problem

This is such a fascinating systems thinking case study because at least 12 key decisions were made by internal Boeing departments and the FAA, each of which does not appear to be a big deal or unsolvable in isolation! Yet, these decisions influenced departmental behavior and interacted with each other to create a disaster-scale complex real-world problem that severely impacted air transportation, not just the internal 737 MAX program at Boeing. There are many interesting details and subtleties that are outside of the scope of this chapter. This section is our attempt to summarize, interpret, and reflect on the material published in these references: [2, 4, 5, 7–9, 11, 12].

We don't want to pick on a particular company. We believe that costly, unanticipated, adverse, emergent behaviors and stories like this can happen to even best-in-class companies with stellar previous track records. We picked this example because we can show that some bad decisions in isolation would not have caused the disaster that emerged from their interplay over time.

3.3.1 Background

Soon after entry into service, the 737 MAX aircraft suffered two fatal crashes: Lion Air Flight 610 from Jakarta to Pangkal Pinang (189 passengers died) and the Ethiopian Air Flight 302 from Bole, Ethiopia, to Nairobi, Kenya (157 passengers died). Both flights suffered an unrecoverable nose-down pitch initiated by the Maneuvering Characteristics Augmentation System (MCAS), an element or a module of the Flight Control Actuation System. In addition to the lives lost, Boeing initially lost $18.4 billion (in fines, settlement, and customer-related fees). The issue reduced the stellar safety reputation of Boeing and tightened FAA regulatory practices impacting the 737 MAX in addition to other new aircraft product lines in development at Boeing and other manufacturers. Unfortunately, this was a twin crisis as it coincided with the COVID-19, reducing Boeing's annual revenue from $100 billion to $58.2 billion.

3.3.2 Factors at Play

Several references describe the background and background factors at play. Here are a few that are regularly referenced:

Competition with Airbus: Airbus introduced the A320neo (neo: new engine option) in 2011 using higher bypass ratio (larger diameter) engines to achieve the lowest $/(Seat X Mile) for that class of aircraft at the time. To compete, Boeing upgraded their 737 design also with a higher bypass ratio turbofan called the 737 MAX. Airbus claimed that pilots with previous A320 experience would not need any simulator or type certification, which added to its attractiveness.

Pitch-Up Tendency Requiring MCAS: Boeing attempted the same upgrade approach to their existing 737. But since the 737 MAX's landing gear was shorter, Boeing had to move the engine forward and upward to accommodate the larger-diameter wings under the wing. This created a pitch-up (i.e., nose-up) tendency during takeoff or at higher angles of attack (AoA). The AoA is the angle of the aircraft's longitudinal axis relative to the straight and level flight path. To compensate for this, Boeing added a Maneuvering Characteristics Augmentation System (MCAS). This used one of two AoA sensors mounted on either side of the cockpit to detect high AoAs and automatically swivel the horizontal tail to lift the tail and push the nose down.

$1 Million Penalty Clause: Southwest Airlines put in a large order for the 737 MAX but added a clause in the purchase agreement that there would be a $1 million penalty per aircraft if the new aircraft required any new training or certification for 737 pilots since, by comparison, the competing A320neo did not require training or certification.

Runaway Nosedive Overspeed Problem: Reliability engineers predicted a catastrophic scenario where the pilot had a maximum opportunity of 10 s to correct a nosedive before the speed was so high that the flight control actuation system did not have enough force to stabilize the aircraft. This was confirmed by test pilots who knew about the existence of MCAS. They concluded that a pilot with good emphairmanship would be able to detect any misbehavior of the MCAS and disable the autopilot within the 10 s allowed.

Decision to Remove MCAS from Operating Manuals: Boeing was presumably confident that there would be no issues and asked the FAA permission to exclude any mention of MCAS in the operating manuals. The FAA gave this permission without noticing any problematic consequences because it deferred to Boeing's expertise. Boeing would probably not have requested this change if they thought it might interplay with the other decisions to create a disaster.

Many Other Factors: Our references cited here point to many other factors that played a role, such as the shifting management culture at Boeing as a result of the merger of McDonnell Douglas with Boeing years earlier, the decision to remove the AoA disagree warning function from some aircraft configurations that was previously a standard function, miscalibrated or out-of-spec AoA sensors, and others.

3.3.3 Interplay of Key Decisions that Lead to Disaster

Figure 3.6 is assembled from the main factors at play. It shows that eight departments (seven internal and one external) made decisions that, from the internal perspective of the departments, did not raise major flags.

The test pilots may have assumed that the 10-second pilot reaction threshold to avert a runaway airplane overspeed was not a big deal under normal circumstances for an experienced pilot. But they were probably surprised by the commercial decision to make the AoA an optional cost versus standard feature on other aircraft. Switching between sensors per flight was also probably something that only a few departments outside of engineering knew about.

We guess that the decision to expand the scope and repurpose the MCAS was made by the project engineering team alone, but this team was also probably not aware of the decision to totally exclude airline pilots from knowing about its existence, removing any reference to it in flight manuals or the $1 million penalty per aircraft if training was needed.

Authorized representatives did not properly inform the FAA of MCAS criticality or concerns, but they may also not have been aware of emergence as a whole versus minor issues that did not appear to be a big deal in isolation.

Sales and marketing also probably did not appreciate the counterproductive impact the $1 million penalty per aircraft clause might have on the behavior of some of the other departments.

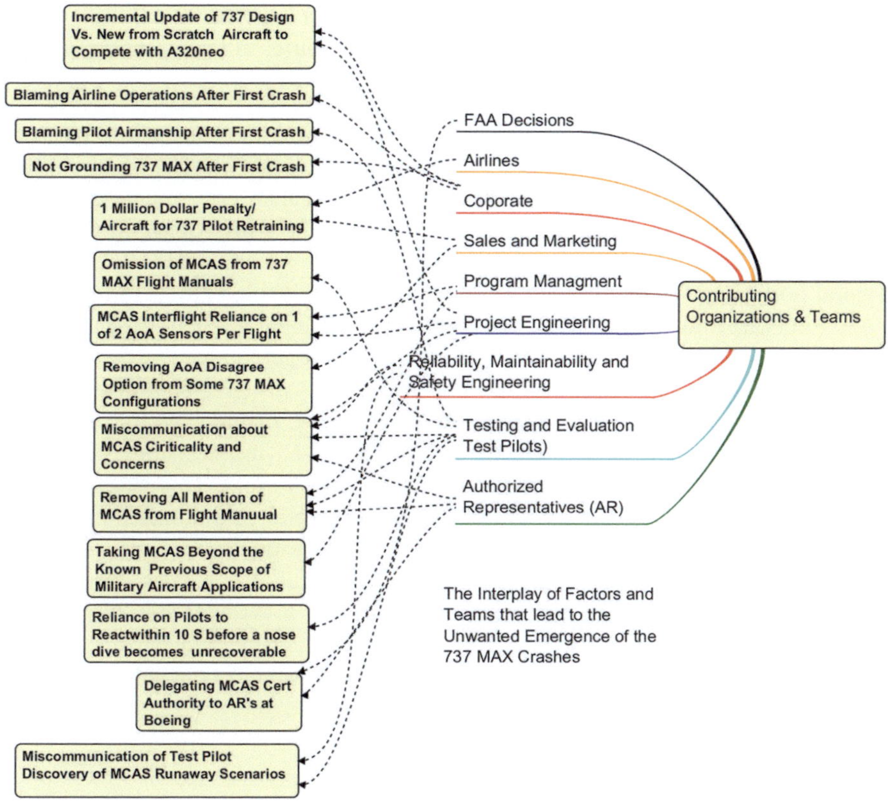

Fig. 3.6 Key decisions that lead to the emergence of the 737 MAX disaster. In isolation, they do not appear to be a big deal. But together, they lead to an air transportation disaster

Figure 3.6 is interesting in that we can remove just one or two of the decisions and possibly avoid the entire unfortunate rollout of the 737 MAX and the huge loss, for example:

Decision to Modify Versus Build New: In the end, the assumption of lower cost and faster time to market of incremental development did not materialize for the 737 MAX. So with hindsight, we can see that the corporate decision to modify versus designing a brand new aircraft turned out to be actually more risky, not less. A new design from scratch would not have required the engines to move to then need a solution to compensate for the pitch-up tendency.

Excluding MCAS from Flight Manuals: Requesting permission from the FAA to exclude MCAS from flight manuals proved to be very shortsighted and damaging in the long run.

Test Pilot Conclusions Based on the Assumption of Good Airmanship: On the one hand, the test pilots experienced MCAS's adverse oscillatory behavior and confirmed the unrecoverable nosedive scenario. On the other hand, their highly reputable chiefs concluded this was harmless because, in this scenario, a good pilot would quickly decide to disable the autopilot.

We expect a regular cadence of sharing and discussing the conclusions between all the stakeholders, which would have probably identified how the isolated decisions in each department can interplay to create a safety-critical emergence.

It is interesting that at the time of writing this text, the 737 MAX is certified to fly again and has not had an MCAS-related failure since. It was grounded for 20 months a good portion of which no doubt was due to the justifiable extra scrutiny by the FAA. Pilots are now required to complete simulator training for their 737 MAX-type rating, and the MCAS existence and behavior are no longer hidden. Furthermore, all 737 MAX production aircraft now have an active AOA disagree alarm function. In the end, the fixes were not hard and easily within the manufacturer's capability. Yet the complex emergence from the interaction of all the decisions in Fig. 3.6 was so hard to predict when it mattered most: before rollout. This is, in essence, similar to the much simpler problem of LEDs icing up in cold climates in Fig. 2.1 that several automotive manufacturers missed because of the way we divide the areas of responsibility and reward people to only focus inside their division and area of responsibility. In an organization that fosters a systems thinking culture, people would ideally balance their efforts and focus inside and outside of their area of responsibility so that they at least understand one or two levels outside the boundaries of the actual deliverable they are responsible for.

3.3.4 Systems Thinking Reflections

The issues that led to the LED icing problem in Section 2.4 and the 737 MAX crisis are common to many industries and organizations. Decisions are made locally or are not transparent to all the relevant stakeholders or people who should be in the know. If you are an LED expert, it is hard to predict the impact of switching to LED to icing the headlamp if your efforts and thinking are limited to the details of the LED and if your responsibility is also limited to the functioning and specification of the LED element alone.

Similarly, suppose we are working for the reliability and safety team for an aircraft manufacturer and we have identified a potential hazard that relies on the pilot to correct in 10 S. In that case, we will certainly speak up if we become aware that the flight manuals removed any mention of the added software module involved. We would shout even louder if we realized that pilots are not required to receive any new training relative to what they already know on previous aircraft that do not have this particular hazard or additional automatic flight control functions.

What if we were the test pilots? Based on our expertise and previous experience, we may be right in concluding that an experienced pilot with good airmanship could quickly determine an issue with the MCAS and disengage the autopilot within

4 seconds. This is comfortably less than the 10-second runaway threshold where the aircraft reaches an unrecoverable airspeed in a nosedive. Yet, suppose we were aware that the AoA disagree alarm only worked for those who paid for an additional AoA disagree alarm option. In that case, it is a very reasonable assumption that we would then raise a flag and insist on new pilot training and the inclusion of the hazard in the flight manual.

What if we were responsible for sales and marketing? Including the hazard in the flight manual and additional training for the pilot is a significant disadvantage relative to our competition, not to mention the $1 million penalty per aircraft. Yet, the engineering team in our company has always proven to be brilliant problem-solvers, and their solutions have worked well for decades. We may even think that if engineering misses something important, the test pilots would surely catch it and not allow the aircraft to enter service. We were probably very happy when the aircraft was released into service, assuming all went well with engineering and testing. We did not realize that those departments were all very busy inside their boxes and unaware of all the decisions in the other camps. At the minimum, at a production rate of, say, 50 per month, we may believe that we just saved the company $600 million dollars a year! But, in fact, the damage in reality was $50 billion, roughly two orders of magnitude higher than the savings we were counting on.

In mid-to-large-sized organizations, we are all divided into separate departments and narrow specialties. We are typically encouraged or rewarded to focus on our narrowly defined areas of responsibilities and deliverables. Unwanted issues, behavior, and problems can often come from the interaction of things we know well with things we do not know well or are not aware of. In essence, the issue is that the individual experts grouped by department are not able to see or predict the impact of their decisions on other experts and departments. They are also so busy with their own deliverables that they either are unaware of or have no time to understand the impact of decisions and changes from other departments. Adopting a systems thinking approach that balances the interactions between technical, human, organizational, and commercial factors is a way to balance short-term compartmentalized gains versus long-term overall concerns and reduce the risk of systemic and systematic failures with significant unwanted consequences.

3.4 Reflection on Chapter 3 Material

We discussed three real-world problems that can benefit from a greater application of systems thinking. The first example is about a social worker tasked with a decision on whether to remove an obese child from his family for health reasons. The second example is the homelessness problem that has exploded in major cities in the USA despite increasing efforts and roughly $3 billion in recurring expenditures to stabilize it. The third example is the Boeing 737 MAX problem where a miscalibrated angle of attack sensor, a very minor and well-understood element of flight controls, interacted with other well-understood elements to cause an *emergence* that costs many lives and more than 50 billion dollars in damage to the manufacturer. All are examples of emergent hard-to-predict behaviors that result from the interplay of several causal relationships that may be well understood in isolation. Here, we emphasized that a lack of systems thinking can have significant real-world consequences. Parts II and III of this book will teach the philosophy and rigorous methods to avoid repeating costly failures in our professional and personal lives.

References

1. Center for Disease Control and Prevention. "Childhood Obesity Facts." n.d. CDC. 3 April 2024. https://www.cdc.gov/obesity/data/childhood.html.
2. Defazio, P. A., & Larsen, R. (2020). Final committee report: The design, development certification of the boeing 737 max. In *House committee on transportation and infrastructure*.
3. Deninno, N. (2011). Obese 200-pound third grader put in Ohio foster care. In: *International Business Times*. https://www.ibtimes.com/obese-200-pound-third-grader-put-ohio-foster-care-375170.
4. Gates, D. (2021). New delay in 777X swells massive Boeing loss in 2020, the largest in its history. In *Seattle Times*.
5. Gates, D., & Baker, M. (2019). The inside story of MCAS: How Boeing's 737 MAX system gained power and lost safeguards. In *Seattle Times*.
6. Harvard School of Public Health. School Obesity Prevention Recommendations: Complete List. n.d. 3 April 2024. https://www.hsph.harvard.edu/obesity-prevention-source/obesity-prevention/schools/school-obesity-prevention-recommendations-read-and-print/#:~:text=School%20Meals%2C%20Competitive%20Foods%2C%20and,schools%20can%20help%20prevent%20obesity.
7. Herkert, J., Borenstein, J., & Miller, K. (2020). The Boeing 737 MAX: lessons for engineering ethics. In *Science of Engineering Ethics*.
8. PBS Frontline Documentary. Boeings Fatal Flaw. (Mar. 2024).
9. Robinson, P. (2021). *Flying blind: The 737 Max Tragedy and the fall of boeing*. Doubleday.
10. Swanson, J. *Systems Thinking!* 30 October 2014. https://www.youtube.com/watch?v=GPW0j2Bo_eY&t=199s
11. Seyer, K., & Londoner, E. (2020). Case study of the Boeing 737 Max 8 crashes using a systems thinking approach. In *Proceedings of the 2020 Annual General Donald R. Keith Memorial Capstone Conference*.
12. Tkacik, M. (2019, October). How Boeing's managerial revolution created the 737 MAX disaster. *The New Republic, 250*(10), 12.

Chapter 4
Introducing Systems Thinking Terms and Concepts Using Man-Made Systems

Abstract The primary intent of this chapter is to get an appreciation for systems thinking terminology, mindset, and the structured approach to real-world complex systems. We present three man-made systems that require systems thinking to understand and influence: air traffic control, the US justice system, and a warfighting system. For each system, we try to better understand the complexity by identifying their relevant systems (context system, system of interest, and enabling systems) and highlight their main: (1) elements, interconnects, and purpose; (2) influence of purpose, structure, and behavior on each other; (3) boundaries of the context system, system of interest (SoI), and enabling systems; (4) life cycle and evolution; (5) interaction and feedback within and between the relevant systems; (6) architectural decisions impacted by mental models; and (7) sensitivity to human reach. This chapter does not try to say that things were not complex before. But it is recognizing that they are getting harder to predict, manage, fund, and execute. Systems thinking is not a magic wand that solves complex systems' problems. It gives us the tools and a framework to understand complex systems and problems arising from systems and possible solutions to solving problems. Creating a systems thinking culture is a key enabler of adopting systems thinking principles, framework, methodology, and tools to solve complex problems.

4.1 Air Traffic Control (ATC)

The air traffic control system is an example of modern engineering and human ingenuity, embodying a complex interplay of technology, expertise, regulation, and adaptability. Its primary goal is to ensure the safety and efficiency of air travel, a task that becomes increasingly challenging with the growing volume of air traffic.

Applying systems thinking to air traffic control offers a robust framework for addressing its complexities and performance. Stakeholders can develop more effective and sustainable solutions by adopting a holistic view, recognizing feedback loops, integrating advanced technologies, and addressing human factors and organizational dynamics. As air traffic continues to grow and evolve, systems thinking

K. Eftekhari Shahroudi et al., *Practical Systems Thinking*,
https://doi.org/10.1007/978-3-031-89394-0_4

will be essential in ensuring that the ATC system can adapt to new challenges and opportunities, maintaining its critical role in ensuring the safety and efficiency of the nation's airspace.

4.1.1 Systems Boundaries: What Are the Relevant Systems (Three Systems) of ATC

To understand complex problems or systems that impact them, we typically need to decide or define the boundaries of a minimum of three systems: context system, system of interest (SoI), and enabling systems. Figure 5.4 in Chapter 5 highlights a simple example of the three systems as related to a person's heart. The SoI is typically what we are responsible for, operating or delivering. So, to be able to define the SoI, we need first to define a particular role, responsibility, and purpose. Let's look at the ATC from the perspective of the ATC chief responsible for the smooth, predictable, and reliable operation and performance of ATC at a major international airport:

System of Interest (SOI): The air traffic control system at an international airport, including all the people (ATC officers) and equipment (computers, screens, means of aircraft tracking, and communication), the ATC tower, and room with human operators, screens, windows, cameras, binoculars, etc. The chief's primary goal is to ensure safe and efficient taxiing to take off, takeoff, land, and taxi to dock, in addition to coordinating air travel within, say, a radius of several hundred miles, including aircraft that are not landing or taking off from the airport where the ATC is located.

Enabling Systems: These are the systems that coexist in the same environment as the SoI but enable, constrain, regulate, or impact its operation in some way. The ATC system operates within a stringent regulatory framework established by national and international aviation authorities, such as the Federal Aviation Administration (FAA) in the USA and the International Civil Aviation Organization (ICAO). These regulations govern all aspects of air traffic management, from the design of airspace and flight procedures to safety standards and emergency protocols. Compliance with these regulations ensures a standardized approach to air traffic control, promoting safety and interoperability across different regions and countries. So, the regulatory system is itself an enabling system for the ATC. Many other enabling systems are outside the boundaries of the SoI: runways, concourses, loading/unloading gates, ATC located at other airports, weather tracking and predicting systems, airline operations crew, pilots, Air Force assets, Department of Homeland Security, cybersecurity defense systems, etc.

Context System: This is the environment and the system that the ATC fits into. The ATC is physically located inside the international airport grounds so the airport is a context system. It is also managed and regulated by the FAA, so in terms of administration, the FAA is a context system. Yet in another sense, the ATC fits into a network of other collaborating ATCs, air transportation, and in some cases military assets.

To better understand complex problems or systems that influence them, systems thinkers often must identify relevant systems and how they interrelate, interact, or influence each other. This means they balance their efforts between looking inward into the SoI and outward into other systems that impact them. This is why we call it system*s* thinking versus the singular systems thinking: systems thinkers realize that a significant source of complexity and potentially unwanted emergent complex behavior can often come from the interaction between what is inside the SoI and what is outside. The YouTube video in Fig. 4.2 is precisely such an example where a software update pushed by CrowdStrike (a computer virus-checking software) created global air transportation chaos, with people stranded at airports all around the world for days. Interestingly, CrowdStrike software is not inside the core hardware and software of ATC. Still, it impacted computers that were controlling transportation to/from airports, check-in systems, airline operation systems, etc., which live in the same environment as the ATC and influenced it in unanticipated ways.

The encouragement to identify the relevant systems to our problem comes from the *3-Systems Principle* explained in detail in Chapter 5. Figure 5.1 highlights a simple example of the *Holism Principle* in making a cake.

4.1.2 Elements, Interconnects, and Purpose

Purpose: From the ATC chief's perspective, the ATC's purpose is to ensure the aircraft's safe and efficient movement within the skies and on the ground. ATC is the system that ensures aircraft do not collide in the air or during landing and takeoff. Figure 4.1 shows the ATC tower where human air traffic controllers are sitting in front of their screens, communicating frequently with pilots to help them accomplish their flight missions. The intricacies of this system stem from the ATC's multifaceted responsibilities, which include managing aircraft during takeoff, landing, and en route phases, coordinating with multiple airports and national air spaces, and adapting to dynamic conditions such as weather changes and unexpected emergencies.

Elements: The main elements of the ATC system are radar systems, communication networks, navigation aids, automation tools in addition to other equipment (hardware, software, binoculars), desks, computers, the control tower, windows, radar, and GPS-based systems that display position and movement of aircraft, ensuring that controllers have real-time data to manage air traffic.

Interconnects: Communication networks enable continuous dialogue between pilots and controllers, facilitating instructions and updates that ensure flight safety and efficiency. Navigation aids, such as GPS and VOR (VHF Omnidirectional Range), guide aircraft along predetermined routes. On board (i.e., in cockpit), navi-

Fig. 4.1 Air traffic control (ATC) in an international airport. Managing ATC is a complex endeavor: weather, delays, maintenance, labor strikes, and geopolitical and health issues thousands of miles away can severely impact its proper operation

gation systems continuously communicate and send aircraft position and trajectory to be for integration with radar-based information to more accurately pinpoint the aircraft. Automation tools, including software for conflict detection and resolution, assist controllers in maintaining safe distances between aircraft. The integration and synchronization of these technologies are paramount, as even minor glitches can have significant repercussions on air traffic management. So, interconnects can be a physical (mechanical or electrical), data link, virtual interaction or the verbal/radio communication line between human controllers and human pilots.

The encouragement to consider, the *elements*, *interconnects*, and *purpose* of a system altogether and not in isolation, comes from the *Holism Principle* explained in detail in Chapter 5.

4.1.3 Influence of Purpose, Structure, and Behavior on Each Other

What ensures that the ATC functions properly for a range of situations like low visibility and bad weather conditions? It is the way the components of the ATC are designed, interconnected, and operated: the *structure*. The structure of the ATC was driven or influenced by the need to safely and efficiently coordinate aircraft in and around airports: the *purpose*. As air traffic volume increased, the required *behavior* and therefore the *purpose* were adapted over time to handle more runways, aircraft,

Fig. 4.2 YouTube video from the July 20, 2024, BBC report on global air transportation chaos resulting from a software that is not an element of air traffic control systems

abnormal scenarios, and higher volumes. As the *purpose* was adapted, so did the *structure* and therefore the *behavior*.

Under normal circumstances, the ATC works very well meaning that the *behavior* is aligned with the *purpose*. In 2023, ATC at Denver International Airport controlled 662,010 aircraft operations and 77 million passengers without any aircraft-related passenger fatalities. ATC is flexible to allow for a number of known issues. For example, if a pilot is late on board a scheduled airline flight because she was stuck in traffic on the way to the airport, then the ATC is informed, the airline flight is rescheduled, and the aircraft position is shifted in the takeoff queue.

Yet, there are situations where the *behavior* of the ATC is not aligned with the stated *purpose* that includes efficiency and fast turnaround times. On July 19, 2024, CrowdStrike, a software company that creates cybersecurity solutions such as virus-checking software, pushed a software update for Microsoft platforms that crippled many critical services worldwide including air transportation for several days stranding millions of passengers. The YouTube video from the BBC in Fig. 4.2 covers the story well. *The New York Times* reported: "On Friday, July 19, alone, nearly 14% of the scheduled flights in the United States were canceled and 56% were delayed." The numbers initially grew worse before getting better as CrowdStrike pushed another software fix.

Although the CrowdStrike problem was probably not anticipated in advance, after the unwanted real-world emergent *behavior* occurs, it provides a motivation to improve or *evolve* the system in retrospect: to adapt the *purpose* and *structure* of the ATC in such a way as to make the *behavior* less sensitive to similar scenarios. So this example shows that *purpose*, *structure*, and *behavior* of systems influence each other all the way throughout the life cycle of the SoI not just during design. This effect is captured in the *Architecture Principle* discussed in Chapter 5. Figure 5.3 highlights a simple example of the Architecture Principle.

4.1.4 Life Cycle and Evolution

ATC needs and required behaviors can evolve due to one or more factors like investments, infrastructure, technologies, routes, travel patterns, real-world issues, safety, and efficiency. This causes the architecture of one or more of the three systems to also evolve so that the emerging behavior of the ATC matches the changing demands.

For example, upgrading the radar system (part of the ATC SoI) not only impacts the human controllers directly but also has implications for training programs, operational procedures, and coordination with other air navigation service providers. By examining the entire system, decision-makers can anticipate these ripple effects and design integrated solutions that address multiple facets of the system simultaneously. This holistic perspective helps ensure that improvements in one area do not inadvertently create problems in another.

Systems thinking can facilitate the integration of new technologies such as NextGen, the FAA's initiative to modernize the US air traffic system, and the increasing use of UAVs. By viewing these technologies as part of a broader system, stakeholders can develop strategies that ensure seamless integration and interoperability. For example, NextGen involves the use of satellite-based navigation and automated data exchange to improve air traffic management. Systems thinking can help coordinate these technological advancements with existing infrastructure, regulatory frameworks, and human factors, ensuring that they complement rather than conflict with each other. The ways of a systems thinkers are using a holistic approach to understand complex systems like regulatory systems and their interrelationships with business and the public. A systems thinker emphasizes seeing the big picture while understanding how individual components interact. The ways of a systems thinkers are to recognize patterns of behavior over time and use Systems Principles to explain or effect change in complex systems. How the ways of a systems thinker can have real-world effect is shown in Section 4.1: Air Traffic Control (ATC).

The *evolution principle*, explained in detail in Chapter 3 reminds us to always consider the life cycle and evolution of systems intended to address real-world problems because a system that is perfectly predictable and knowable can become very hard to understand or anticipate as it evolves and changes form throughout its life cycle. As systems within the SOI such as the radar system get outdated and eventually upgraded, the airport (context) and enabling systems such as airline operations will be impacted. This is part of the life cycle of the ATC which will evolve as the constituent elements get upgraded.

4.1.5 Interaction and Feedback Within and Between the Three Systems

It's curious that a small piece of CrowdStrike software that is not used in aircraft and any of the critical subsystems of the ATC (such as radar, radio, and GPS) would be able to cripple air transportation on multiple continents for several days. Since passengers and crew are required to arrive well in advance of scheduled flights, then when the software was pushed, many of the aircraft, crew, and passenger were already at the airport. So how come so many were cancelled and delayed? The answer is that unwanted emergent behavior is often the result of interactions of the SOI with enabling and context systems, so the cause is often not purely internal to the SoI. Even if the SOI is not damaged or compromised internally, it still relies on inputs and communications from many other systems outside of it.

The check-in computers (part of an enabling system outside the SoI) help airline crew to scan passenger IDs, print boarding passes, print luggage tags, weight luggage, and accept luggage. If they were using Windows 10 or 11 software and got the CrowdStrike update, they would hang up. So the incoming flow of passengers needing to check in quickly piles up since the airline is unable to accept any luggage when their computers are down. There is not much the airline crew can do to relieve the situation: their work process can handle a small number of computers crashing, but they appear to have no recourse when all their computers hang up. Taking off with a significant portion of passengers not reaching the gate is also not a great option, as the airline is legally liable for passenger losses. Now, consider aircraft that land and taxi to the gates to unload their passengers, but all the spots are taken by the aircraft that are delayed and were not able to take off. So, for the ATC that is our SoI to be severely impacted, it is not necessary for the cause to be a malfunction of one of its internal parts. It can be an issue that comes about through interaction and feedback between the SoI, the context system, and enabling systems.

Another example is the gate computers used to check boarding passes and IDs prior to allowing passengers to board. For both safety and security reasons, Federal Aviation Regulations (FAR) Section 121.693 requires that all certificate holders prepare a load manifest that includes, at the time of takeoff, the names of passengers. So, even if the pilot, aircraft, and passengers are ready, the inability to smoothly check boarding passes, board the aircraft, and communicate the flight manifest to ATC to get permission to take off is compromised when the gate computers crash. Again, this is an example of interaction and feedback between the three systems and not an issue caused by an internal malfunction of a part of the ATC.

So, when we try to understand or anticipate hard-to-predict real-world complex issues that impact critical systems like the ATC, our chances improve if we do not purely focus inside the ATC. Instead, systems thinking encourages us to balance our efforts and focus to also think about the interaction between the relevant context systems (airport, FAA, US ATC that is a network of local ATCs), enabling systems (check-in, ground crew, airline operations and maintenance, etc.), and the SoI (ATC). Interactions within and between the three systems often cause the complex emergent behaviors we are trying to understand and influence. This is captured in the *Emergence Principle* explained in detail in Chapter 5.

Architectural Decisions Impacted by Mental Models

Air traffic control centers started operating in the USA in the 1930s to first coordinate landings and takeoff. Soon, their role grew to coordinate aircraft on route. The dominant overarching mental model is that you cannot purely rely on the equipment and flight crew on board the individual aircraft to guarantee safe and efficient air transportation. This high-level mental model has persisted and incrementally adapted as new technologies at the ATC and on board the aircraft become available. Safe, reliable, and efficient air transportation is essentially still very dependent on human operators in the ATC in frequent communication with human pilots. If we look at the high-level architecture and operation of the ATC, we see that, over time, it has incrementally grown to support higher traffic volumes: larger towers, more screens, more human operators, more accurate and redundant radars, more communication and overlap between individual ATCs, addition of GPS-based systems, live weather updates, etc. So, everything has grown into a much larger-scale version of what was started in the 1930s with improved components, but the high-level purpose, architecture, and operation of the ATC have essentially remained similar because our mental model of transporting people and cargo has remained the same.

However, our mental models are adaptable as our needs, technologies, and societies evolve. Changes to our mental models may not always be just incremental adaptations. We may, for example, start trusting fully autonomous aircraft that do not require human pilots or a flight crew. This would be a significant abrupt shift in our mental models about air transportation where the current architecture of the three systems are no longer capable of producing the required behaviors. Furthermore, just scaling up the existing ATC architecture would not be practical. An increase of air transportation by a factor 100 by adding fully autonomous smaller aircraft would probably shift a significant portion of the ATC functions to smarter networked interconnected systems on board the flying aircraft. Suddenly, a large portion of FAA regulations will require changing, so, in the end, an abrupt change to our mental models will eventually drive significant changes to the architecture of all three systems, not just the SoI. This concept of the adaptability of our mental models and its impact on human decision and system architectures is captured in the *Architecture Principle* and discussed in detail in Chapter 5.

Sensitivity to Human Reach

The complexity of ATC is a result of the interplay between advanced technology, human expertise, stringent regulations, and the unpredictable nature of air travel. Despite the reliance on technology, human expertise and decision-making remain at the core of the ATC system. Air traffic controllers are highly trained professionals who undergo rigorous training and continuous assessment to maintain their skills. They are responsible for making real-time decisions based on complex information, requiring a deep understanding of aviation regulations, aircraft performance, and meteorological conditions. Controllers must coordinate with multiple parties, including pilots, ground staff, and other control centers, to seamlessly manage air traffic flow. The cognitive demands of the job are immense, as controllers must process vast amounts of data quickly and accurately while maintaining a high level of concentration and composure. For example, comprehensive training programs that incorporate systems thinking principles can better prepare controllers to handle complex and dynamic situations. Some ATC programs may have systems thinking when environmental and personnel factors are considered. But when considering the entire air traffic transportation system, ATCs are usually focused on their area of influence. Additionally, fostering a collaborative organizational culture can enhance communication and coordination among controllers, pilots, and other stakeholders.

Even if our mental models change in the future and we decide to make the ATC and flying aircraft fully autonomous as explored in the last section, humans still need to be a major consideration for all the stakeholders involved in this change, because human mental models and behaviors are still able to reach and interact with the three systems to cause difficult-to-predict and hard-to-mitigate unwanted emergent complex behaviors. This effect is captured by the *Human Principle* explained in detail in Chapter 5.

4.1.6 Sources of Complexity

ATC has a safety critical role and faces numerous challenges, including aging infrastructure, increasing traffic volumes, and the integration of new technologies such as unmanned aerial vehicles (UAVs) and commercial space flights. There are days where the ATC works like a finely tuned clockwork and appears to be very predictable. On those days, you may be forgiven for thinking that the ATC is not a complex system or that air transportation is not a complex hard-to-predict and hard-to-mitigate problem. Yet, as the BBC report in Fig. 4.2 showed, it did not take much, a small software update, to severely disrupt ATC on multiple continents at once!

Sources of complexity can be many: for example, evolving rules and regulations from the FDA add layers of complexity, as controllers must stay abreast of changing rules to ensure their actions align with legal requirements. The FDA-mandated requirement to communicate a passenger list to the ATC prior to getting permission for takeoff in the flight is normally not a big deal, but when this requirement interplays with crashed computers at the gate, it stops the flight and creates a pile up of aircraft arriving at the gate compounding delays. Weather conditions, such as thunderstorms, fog, and turbulence, can significantly impact flight operations, requiring controllers to adjust flight paths and schedules on the fly, as well as unexpected emergencies, such as mechanical failures or medical incidents on board.

There are many more sources of complexity impacting the ATC including the fact that it relies heavily on human operators inside and outside of the ATC, interconnected operation of national and international ATCs, etc.

Similar to the 737 MAX example of Fig. 3.6 and the CrowdStrike example of Fig. 4.2, the sources of complexity by themselves may not be a huge deal, but through the interactions of the three systems, they can easily add up to create major unwanted emergent behavior that is hard to mitigate or manage when it occurs. Unfortunately, too many people have learned to look, think, and act purely inside their narrow area of responsibility and not wonder if our systems can be better by also looking at how the parts and systems interact and influence each other.

In trying to improve our systems that deal with real-world problems, systems thinking encourages us to identify all the relevant sources of complexity and understand the potential ways they can reach, influence, and amplify to cause unwanted emergence. The *Systems Principles* of Chapter 5 guide our thinking, actions, behaviors, and methods to improve our chances of understanding how complex unwanted behavior emerges and how to reduce their impact.

4.2 Justice System

Systems thinking encourages us to understand the how and why of the way the whole behaves and in what way the system is more than just a list of components:

The justice system is an example of a complex, nontechnical system that has a large effect on society. The purpose of highlighting the justice system shown in Fig. 4.3 in systems thinking is because it demonstrates how interconnected components work together to achieve fairness, accountability, and societal stability if the entire system is considered as a whole. The justice system is a complex entity that functions not only through its individual components, courts, law enforcement, corrections, and legislation but also through the interplay of these parts. It embodies the principles of fairness, equality, and morality, transcending its structural elements to uphold societal order and protect citizens' rights. The system's strength lies in its collective ability to adapt and respond to the evolving needs of society, ensuring that justice is not merely a concept but a living, breathing manifestation of our collective values and aspirations.

4.2.1 Elements, Interconnects, and Purpose

Systems thinking promotes a holistic understanding of the justice system, recognizing it as an intricate network of interrelated components, including police departments, courts, and correctional facilities.

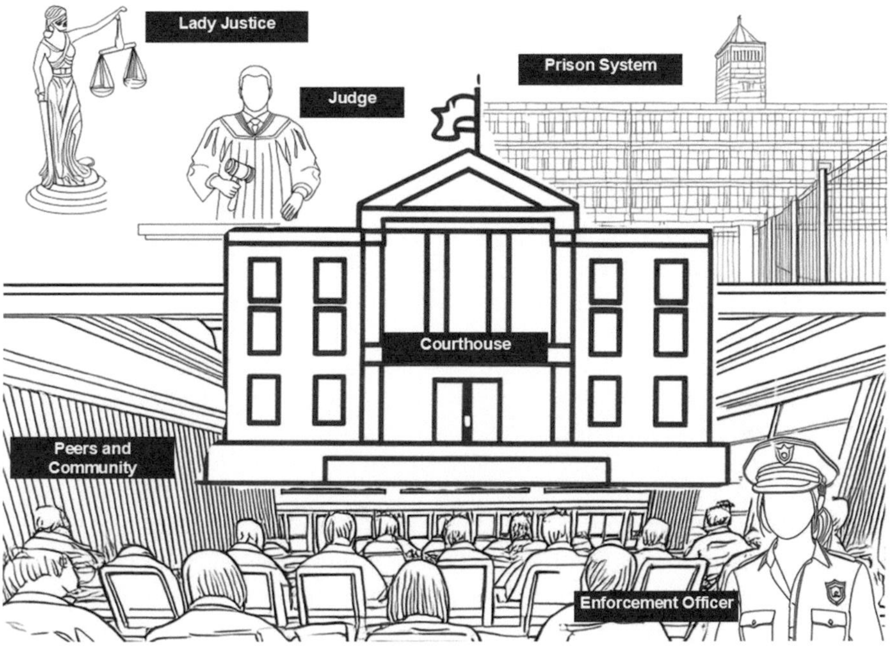

Fig. 4.3 The US justice system of interconnected elements: courthouses, courtrooms, law enforcement, and various other branches

4.2.2 *Influence of Purpose, Structure, and Behavior on Each Other*

Systems thinking shows us that the purpose, structure, and behavior of the US justice system are intricately interwoven within the framework of the rule of law, each influencing and reinforcing the other. The system's purpose is to maintain social order, protect individual rights, and provide a mechanism for conflict resolution, which is reflected in its structure that includes a network of courts, law enforcement agencies, and correctional institutions. This structure is designed to administer justice fairly and efficiently, guided by the Constitution, which ensures that laws are applied consistently and impartially. The behavior of the justice system, through its agents and officers, is expected to embody the principles of equality, fairness, and respect for the rule of law, which in turn shapes the public's perception and trust in the system's purpose and structure. Due to the diversity of the agents that are tasked with caring out the rules of law, unwanted behaviors can emerge in which the designed system of justice behaves in an unjust or fair manner. Systems thinking can be used to understand occurrences of such emergent behavior. The rule of law is supposed to act as the bedrock of the US justice system, ensuring that the purpose is aligned with fair equitable values, the structure is upheld by legal frameworks, and the behavior is supposed to be regulated by ethical and legal standards. Thus, the US justice system operates as a cohesive entity, with its purpose, structure, and behavior dynamically interacting to uphold the law and deliver justice.

Systems thinking allows us to examine the true purpose of a system which is the biggest determinant of structure, behavior, and outcomes.

The true purpose of the justice system is to maintain social order by enforcing laws, ensuring fairness, and resolving disputes. It aims to protect the rights of individuals and uphold the rule of law. The biggest determinant of its structure, behavior, and outcomes is the legal framework within which it operates, which is shaped by the Constitution, legislation, and societal values. This framework influences how justice is administered and the effectiveness of the outcomes in reflecting the principles of equity and justice.

Purpose: The US justice system serves a multifaceted role within society, primarily aimed at maintaining law and order, protecting individual rights, and ensuring justice is administered fairly and impartially. Its foundational purpose is to interpret and apply laws through a structured court system, where disputes are resolved and legal judgments are made. The justice system is also responsible for upholding the Constitution, including the protection of freedoms and liberties guaranteed to all citizens. It seeks to deter criminal behavior by imposing penalties on those who violate laws while also providing a mechanism for rehabilitation. Furthermore, the system functions to balance competing interests, mediate public and private disputes, and facilitate the peaceful resolution of conflicts. Through these actions, the justice system strives to preserve social harmony, safeguard democracy, and uphold the rule of law, which are essential for a stable and just society.

Structure: The US justice system is a complex web of legal frameworks designed to uphold the law, administer justice, and protect the rights of individuals. At its core, the system is divided into three main branches: federal, state, and local. The federal level is headed by the US Supreme Court, which has the ultimate authority on constitutional matters. Below the Supreme Court are the US Courts of Appeals and the US District Courts. State justice systems have their own courts, including supreme courts, appellate courts, and trial courts, which handle civil and criminal cases under state law. Local courts, such as municipal and county courts, deal with lesser civil disputes and misdemeanors. Each level operates within a framework of checks and balances, ensuring no single entity holds absolute power. The justice system also includes law enforcement agencies, which enforce laws and maintain public order, and correctional institutions, which manage the punishment and rehabilitation of convicted individuals. The system is governed by a body of laws, including the Constitution, statutes, regulations, and case law, which guide legal proceedings and ensure fair treatment. The adversarial process, where two opposing sides present their cases to an impartial judge or jury, is a cornerstone of the US justice system, promoting fairness and accountability. Additionally, the system provides mechanisms for legal representation, appeals, and the protection of individual rights, reflecting the democratic principles upon which the nation was founded.

Behavior: The US justice system's approach to sentencing has been a subject of extensive debate and reform. Historically, the system has been criticized for its harsh sentencing policies, which have contributed to the phenomenon of mass incarceration. According to a report by the Vera Institute of Justice, there is a growing consensus that severe sentences do not effectively deter crime, nor do they aid in the healing of crime survivors. The report suggests that the US sentencing system often overestimates the danger an individual poses to the community and when incarceration is necessary for public safety. The Sentencing Reform Act of 1984 marked a significant shift in federal sentencing, introducing guidelines aimed at increasing uniformity and severity in sentencing. However, this has led to mandatory minimums and long-term sentences, particularly impacting Black and Latino communities, who are disproportionately represented in prison populations. The US Sentencing Commission provides an overview of the current federal sentencing process, which includes considering factors such as the nature and circumstances of the offense, history and characteristics of the defendant, and the need for the sentence to serve various purposes like deterrence, protection of the public, and rehabilitation of the offender. There is an ongoing discussion about the need for a paradigm shift towards a system that prioritizes liberty, emphasizes restorative justice, and reserves incarceration for the most serious offenses, potentially reducing mass incarceration significantly. The realities of sentencing in the US justice system reflect a complex interplay of historical policies, current practices, and evolving perspectives on justice and rehabilitation.

4.2.3 Systems Boundaries

The US justice system is delineated by a complex set of boundaries that define its jurisdiction, authority, and operational scope. At its core, the system is structured across various levels, including federal, state, and local jurisdictions, each with its own set of laws and courts. The federal justice system handles cases that cross state lines or involve federal laws, while state systems manage cases within their own borders based on state legislation. The Constitution acts as the supreme legal boundary, with the Bill of Rights providing fundamental protections to individuals, ensuring that the justice system operates within the confines of what is fair and just. Moreover, the separation of powers creates checks and balances between the legislative, executive, and judicial branches, preventing any one branch from overstepping its boundaries. The system also recognizes the boundaries of legal practice, where only those who are licensed and meet certain criteria can practice law. Additionally, the justice system is bound by procedural rules, such as the rules of evidence, which govern what information is admissible in court. Boundaries are different from viewpoints of diverse entities with in the US justice system. A judge views the justice system as a complex framework designed to balance the scales of fairness and order. They see themselves as guardians of the law, tasked with interpreting statutes and precedents with impartiality, while also considering the societal and moral implications of their rulings. For a police officer, the justice system represents a procedural guide for maintaining public safety, where each step—from investigation to arrest—must align with legal standards to ensure the integrity of the process. The boundaries are clear: uphold the law, protect the community, and respect the rights of all individuals involved. Conversely, an individual accused of a crime might perceive these boundaries as opaque or even arbitrary, feeling caught in a labyrinth where power and procedure can seem overwhelming and justice feels distant. Their interaction with the justice system is often fraught with anxiety and uncertainty, and the outcome can profoundly impact their view of fairness and their place within society. Each perspective is shaped by distinct roles and experiences, highlighting the multifaceted nature of justice and its reach in human lives.

4.2.4 Interaction and Feedback Within and Between the Three Systems

Systems thinking explains that systems come in at least three: system context, system of interest, and enabling system.

The justice system can be understood through the lens of three interconnected systems: the system context, the system of interest (SOI), and the enabling system. The system context encompasses the broader legal environment and societal norms within which the justice system operates, defining its scope and limitations. The SOI is the justice system itself, with its network of courts, law enforcement agencies, and correctional institutions working cohesively to enforce laws and deliver justice. Lastly, the enabling system includes the legislative bodies and educational institutions that establish the legal framework and cultivate the professionals who uphold the system's integrity, ensuring its effective functioning and evolution over time. Together, these systems interact to form a complex, adaptive mechanism for maintaining social order and protecting citizens' rights.

System of Interest (SOI): The US justice system is a complex network designed to uphold the law, administer justice, and protect the rights of individuals. At its core, the system is structured around the Constitution, which provides the legal framework for its operation. The system is divided into various levels, including federal, state, and local jurisdictions, each with its own set of laws and courts. The federal justice system handles cases that involve federal law or constitutional issues, while state and local systems deal with matters pertaining to state laws and community issues.

Enabling System: The US justice system is underpinned by a complex enabling system that ensures its functionality and adherence to the rule of law. This system is characterized by a set of principles and structures that include the Constitution, which provides the legal framework; the Supreme Court, which interprets laws; and lower federal courts, which apply laws to individual cases. Additionally, the justice system is supported by various law enforcement agencies that enforce laws and correctional institutions that manage the consequences of crime. A key aspect of this enabling system is the separation of powers, which ensures a balance between the legislative, executive, and judicial branches of government. Moreover, the federal nature of the USA allows for a diverse range of justice systems at the state and local levels, each tailored to their specific needs and challenges. The system also includes numerous actors such as judges, lawyers, and juries, who play critical roles in the administration of justice. Furthermore, recent reforms like the First Step Act aim to address issues such as mandatory minimum sentences and improve conditions within the federal prison system. Overall, the enabling system of the US justice system is designed to maintain social order, protect individual rights, and provide a mechanism for resolving disputes.

Context System: The system reflects the nation's historical and social complexities, including the challenges of addressing racial disparities and ensuring equal justice for all citizens. As society evolves, so does the justice system, adapting to new social norms, technological advancements, and understandings of criminal behavior to better serve the public and uphold the rule of law.

Architectural Decisions Impacted by Mental Models

There are multiple layers of jurisdiction, authority, and responsibility that span state, county, municipal, and federal boundaries. These layers employ thousands of individuals in roles such as judges, commissioners, clerks, bailiffs, and advocates. The criminal justice system in the USA has evolved from English common law into a complex series of procedures and decisions within a system with three levels of government participating: federal, state, and local. This three-tiered system adds to the complexity, as each level has its own set of laws and procedures. The criminal justice system in the USA consists of law enforcement, the court system, and corrections. These three subsystems encompass private and government agencies at the state, federal, and local levels, all of which work together to maintain public safety. Each subsystem handles a different responsibility, adding to the complexity.

Sensitivity to Human Reach

Equity and fairness are central to a just legal system. Systems thinking can help address disparities by examining how different components of the justice system impact various demographic groups. Racial and socioeconomic biases, for example, can be perpetuated through seemingly neutral policies that disproportionately affect marginalized communities. By using systems thinking to analyze data on arrests, sentencing, and incarceration, policymakers can identify and address patterns of discrimination. This approach can inform the development of policies that promote equity, such as diversion programs for nonviolent offenders and initiatives to ensure equal access to legal representation

4.2.5 Sources of Complexity

The justice system, a complex network of institutions, laws, and actors, operates on the principle that its collective function transcends the capabilities of its individual components. It's an intricate ecosystem where law enforcement, judiciary, corrections, and community services work in concert to uphold the rule of law, protect rights, and maintain social order. Each segment, from the police officers who enforce laws to the judges who interpret them, from the lawyers who advocate for justice to the correctional facilities that rehabilitate, contributes a unique and vital role. Together, they embody the ideals of fairness, equity, and due process, ensuring that justice is not merely a concept but a living, breathing manifestation of societal values. This synergy creates a dynamic where the whole becomes greater than the sum of its parts, reflecting a holistic approach to justice that seeks not only to punish but also to educate, rehabilitate, and deter, fostering a safer and more just society for all.

Understanding the court system in the USA is no easy task. There are multiple layers of jurisdiction, authority, and responsibility that span state, county, municipal, and federal boundaries. These layers employ thousands of individuals in roles such as judges, commissioners, clerks, bailiffs, and advocates. The criminal justice system in the USA has evolved from English common law into a complex series of procedures and decisions within a system with three levels of government participating: federal, state, and local. This three-tiered system adds to the complexity, as each level has its own set of laws and procedures. The criminal justice system in the USA consists of law enforcement, the court system, and corrections. These three subsystems encompass private and government agencies at the state, federal, and local levels, all of which work together to maintain public safety. Each subsystem handles a different responsibility, adding to the complexity.

4.2.6 Life Cycle and Evolution

Systems thinking recognizes that systems have a life cycle and they evolve. This is highlighted by the Evolution Principle described in Chapter 5. Systems thinking provides an understanding of the life cycles and evolution of complex systems and elements within the three systems (SOI, context, and enabling systems) which affect each other often causing a ripple effect over time. For example, the evolution of videoconferencing (enabling system) due to the advances in technology has impacted how pre-trial hearings and arraignments in many states are conducted. In this case, the evolution of an enabling system has caused a change in the SOI systems of local and federal laws which has led to reductions in cost and improvements in efficiency of the context system.

The justice system has undergone a significant evolution from its ancient origins to the present day. Initially, justice was dispensed through severe physical punishments and exile, reflecting the harsh realities of early societies. Over time, the influence of the Catholic Church and Enlightenment thinkers led to more humane approaches, emphasizing rights and proportionate punishment. The Industrial Revolution further transformed the system, introducing professional police forces and the concept of rehabilitation through probation and parole. Today's justice system continues to evolve, integrating technology and progressive reforms to meet contemporary societal needs.

The complexity of the US justice system often results in inefficiencies and inconsistencies. Systems thinking can help manage this complexity by simplifying processes and creating adaptable structures. For instance, the proliferation of legal codes and regulations can create confusion and unequal application of the law. By focusing on core ST principles such as the 3-Systems and Boundaries Principles described in Chapter 5, one can identify the relationships and feedback loops between systems and use the Model-Based Systems Thinking (MBST) framework, methodology and tools listed in Chapters 11 and 12 to streamline legal processes and change the culture to support this modification of processes as described in Chapter 9. This will enable the justice system to become more transparent and accessible. Additionally, adaptability is crucial in a dynamic society; systems thinking encourages continuous assessment and evolution of policies to respond to emerging challenges such as new forms of crime and technological advancements.

Equity and fairness are central to a just legal system. Systems thinking can help address disparities by examining how different components of the justice system impact various demographic groups. Racial and socioeconomic biases, for example, can be perpetuated through seemingly neutral policies that disproportionately affect marginalized communities. By using systems thinking to analyze data on arrests, sentencing, and incarceration, policymakers can identify and address patterns of discrimination. In most cases, this is done. In the case of drug enforcement, data show that people of color are observed for longer periods of time than others.

4.3 Warfighting System

Figure 4.4. shows the main elements of a modern warfare system.

4.3.1 Elements, Interconnects, and Purpose

Systems thinking is essential in modern warfighting. Due to technological advances, modern warfighting is now a complex system where purpose, structure, and behavior are intricately linked, each influencing the other within the established rules of engagement. The complexity of modern-day warfighting comes from our world being so much more interconnected today due to technology. The feedback loops occur so much faster than human decision-making can keep up with. Because of the near instant feedback of connectivity, warfighting exhibits complexity in

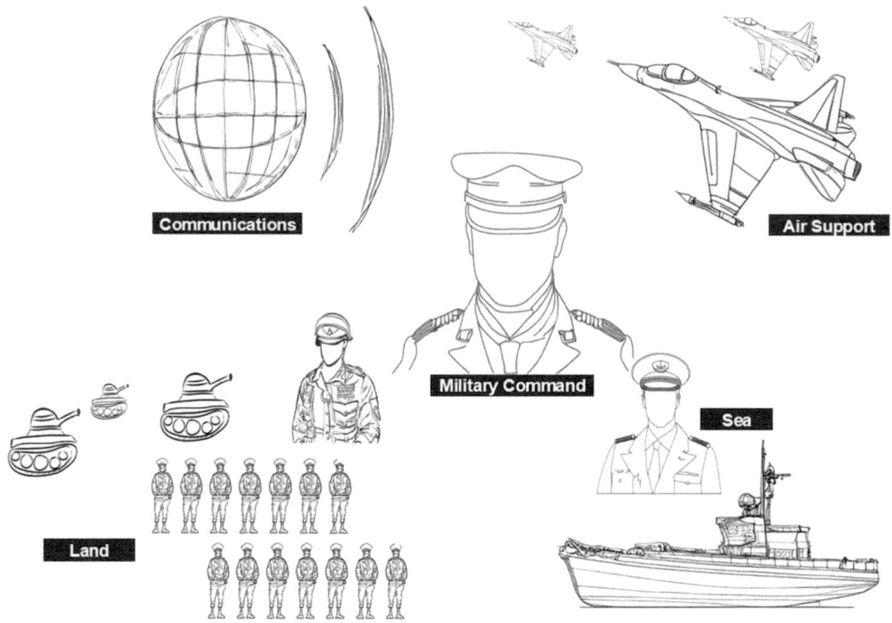

Fig. 4.4 Here is the overview image of modern warfare as a system. It includes elements such as a command center, soldiers in advanced combat gear, drones, military vehicles, and cyber warfare components, capturing the complexity and integration of technology in contemporary military operations

modern times, making it more important to take an ST perspective. The purpose, often defined by political and strategic objectives, dictates the overarching goals of military operations. This purpose drives the structure, which encompasses the organization of military forces, the hierarchy of command, and the allocation of resources, ensuring that the conduct of war is systematic and coordinated. Behavior in warfighting refers to the tactics, strategies, and actions employed by military personnel, which are shaped by the purpose and enabled by the structure. These behaviors must adhere to the rules of battle, which are designed to regulate conduct, minimize unnecessary suffering, and maintain discipline. The rules of battle, in turn, influence the purpose by setting boundaries on what can be achieved through military means. They also impact the structure by necessitating a framework that can operate within these legal and ethical constraints.

Thus, in modern warfighting, the purpose provides direction, the structure offers a means to achieve that direction, and behavior demonstrates the execution, all within the confines of the rules of battle, creating a dynamic interplay that evolves with the changing nature of warfare. This system is designed to be adaptive, allowing for rapid response to threats and challenges while maintaining the integrity and effectiveness of military operations.

Systems thinking encourages us to understand the how and why the whole of a system behaves due to the interaction of parts and in what way the system is more than just a list of components, elements, and interconnects.

Modern warfare is a complex system where various elements such as advanced technology, strategic doctrines, and human elements interconnect to create a dynamic and multifaceted environment. These components work in concert to fulfill a broader purpose, which often extends beyond immediate military objectives to encompass political, economic, and social goals. The interplay between these parts creates a system with emergent properties that cannot be understood by examining the components individually, thus making modern warfare more than the sum of its parts. Modern warfare represent a complex integration of various components that, when combined, create a multifaceted mechanism far different from the mere sum of its parts. This holistic approach to warfare encompasses advanced technology, strategic doctrines, and coordinated operations that leverage the strengths of each component to achieve a synergistic effect. For instance, the integration of cyber capabilities with traditional military assets allows for operations that can disrupt enemy communications and command systems without physical confrontation.

Similarly, the use of unmanned systems and artificial intelligence in surveillance and reconnaissance missions enhances situational awareness and decision-making processes, leading to more precise and effective engagements. The fusion of human skills with technological advancements, such as augmented reality in training and operations, exemplifies how modern warfare systems are evolving into intricate networks that are more adaptive, resilient, and capable than their individual elements alone. This paradigm shift underscores the importance of a comprehensive approach to security and defense, where the interplay between technology, human expertise, and strategic innovation is critical to maintaining superiority in a rapidly changing battlefield environment.

Modern warfare has evolved significantly over the years, becoming increasingly complex due to a variety of factors. Hybrid warfare has gained significant relevance in recent years as states employ non-state actors and information technology to subdue their adversaries during or in the absence of a direct armed conflict. Hybrid warfare entails an interplay or fusion of conventional as well as unconventional instruments of power and tools of subversion.

4.3.2 Influence of Purpose, Structure, and Behavior on Each Other

Systems thinking enables us to examine the true purpose of a system as its true purpose is the biggest determinant of structure, behavior, and outcomes.

The true purpose of a modern warfare system, often rooted in the strategic objectives of national security and geopolitical interests, significantly influences its structure, behavior, and outcomes. The design and organization of military forces (structure) are tailored to fulfill specific strategic roles, whether defensive or offensive. This, in turn, shapes the conduct (behavior) of military operations, dictating tactics and engagement rules. Ultimately, these factors converge to determine the effectiveness and consequences (outcomes) of military engagements, impacting not only the immediate conflict but also the broader geopolitical landscape.

Purpose: The US military serves a critical role in the defense and security of the nation. Its primary mission is to protect the country's sovereignty and defend its interests both domestically and internationally. The military ensures readiness to deter war and, if necessary, to fight and win against adversaries. It supports national policies and implements national objectives while also preserving peace and security. The various branches of the military, including the Army, Navy, Air Force, Marine Corps, Space Force, and Coast Guard, each have specific functions and domains of operation, but collectively they work to maintain the freedom and safety that the USA upholds. The military also contributes to the unity and identity of the nation, having played a decisive role throughout the country's history.

Structure: The US military is a complex and hierarchical structure designed to provide national defense and support military operations. It is composed of six service branches: the Army, Marine Corps, Navy, Air Force, Space Force, and Coast Guard. Each branch has a distinct role, with the Army responsible for land-based operations, the Navy for maritime activities, the Marine Corps specializing in amphibious warfare, the Air Force overseeing air and space superiority, and the Space Force dedicated to space warfare. The Coast Guard, while also a maritime force, is unique in its peacetime missions that include maritime law enforcement and search and rescue. These branches are organized under the Department of Defense, except for the Coast Guard, which is under the Department of Homeland Security during peacetime but can be transferred to the Department of the Navy during wartime. The President of the USA is the Commander in Chief, leading the military with the advice of the Secretary of Defense and the Joint Chiefs of Staff, which includes the highest-ranking military officers from each branch. The military also encompasses a vast reserve component, including the Army National Guard and the Army Reserve, which can be called upon in times of need. The structure is designed to ensure readiness, flexibility, and the ability to respond to various threats and challenges, both domestic and international.

Behavior: The US military operates under a strict code of conduct and a set of core values that guide the behavior of its personnel. These values include loyalty, duty, respect, selfless service, honor, integrity, and personal courage. The military culture is built on the foundation of these principles, aiming to foster a disciplined and effective fighting force. The behavior of military members is expected to align with these values, both in and out of uniform, which is essential for maintaining the trust and confidence of the American public and allied nations. The military also emphasizes the importance of teamwork, leadership, and the continuous development of skills and knowledge to adapt to the ever-changing landscape of defense and security. Additionally, the US military is subject to civilian control, ensuring that its actions remain in line with the democratic values and policies set forth by elected officials. This structure is designed to prevent any single entity from wielding unchecked power and to keep the focus on national security objectives. The behavior of the US military is also influenced by international laws and treaties, such as the Geneva Conventions, which dictate the conduct of armed forces during wartime, including the treatment of prisoners of war and civilians.

4.3.3 Systems Boundaries

The boundaries of the US military are not rigidly defined but are conceptualized through the levels of warfare—strategic, operational, and tactical—which serve as frameworks for organizing military efforts and linking tactical actions to the achievement of national objectives. These levels help commanders design and synchronize operations, allocate resources, and assign tasks to the appropriate command. The strategic level concerns itself with national policy and military strategy, often involving high-level decision-making and long-term planning to achieve political objectives. The operational level bridges the gap between strategy and tactics, focusing on campaigns and major operations to arrange tactical engagements in a manner that supports strategic goals. The tactical level deals with the actual conduct of combat operations and the execution of plans on the battlefield. While these levels provide structure, they are interconnected and fluid, allowing for flexibility and adaptability in response to dynamic operational environments. The systemic boundaries of warfighting vary from different perspectives.

Modern warfighting is characterized by a complex interplay of strategic, operational, and tactical elements that transcend traditional boundaries. From a policymaker's perspective, it involves the careful balancing of military objectives with political goals, ensuring that the use of force is aligned with national interests and international laws. A flag officer, on the other hand, must navigate the operational art, integrating new technologies and capabilities into joint operations to achieve strategic objectives while maintaining agility and coherence across diverse military domains. The soldier's viewpoint is grounded in the tactical reality of the battlefield, where the immediacy of combat demands adaptability, resilience, and an understanding of how strategic decisions translate into direct action. This multifaceted approach to modern warfighting reflects the need for a unified model that accommodates the rapid pace of technological change and the evolving nature of global threats. As warfare becomes increasingly networked and multi-domain, the boundaries between peace and conflict blur, requiring all levels of military engagement to operate with a heightened sense of situational awareness and a robust framework for decision-making.

4.3.4 Life Cycle and Evolution

The US military has undergone a significant evolution since its inception during the American Revolutionary War. Initially formed in 1775 with the establishment of the Continental Army, Navy, and Marines, the US military was created to fight for independence from Great Britain. This period marked the birth of a military force that would grow and adapt through various conflicts and societal changes. The early military was a blend of militia forces and a regular army, reflecting the ideological debate between the need for a citizen-soldier model and a professional standing Army. As the nation expanded, so did the military, playing crucial roles in the War of 1812, the Mexican-American War, and the Civil War, each conflict shaping its structure and strategy. The twentieth century saw the US military's transformation into a global superpower, particularly after World War II. The Cold War era prompted further growth and technological advancement, leading to the establishment of the Air Force as a separate branch and later the Space Force. Today, the US military is a complex organization with a life cycle that includes recruitment, training, active service, and veterans' affairs, ensuring a continuous process of renewal and adaptation. This life cycle is essential for maintaining the military's readiness and effectiveness in protecting the nation's interests at home and abroad.

4.3.5 Interaction and Feedback Within and Between the Three Systems

Systems thinking explains that systems come in at least three: system context, system of interest, and enabling system.

In modern warfare, systems are often conceptualized in a triad structure for clarity and functionality. The system context refers to the broader environment in which the system of interest operates, including external factors and influences. The system of interest is the primary system under consideration, designed to fulfill specific mission objectives. Lastly, the enabling system includes all supporting systems and processes that ensure the system of interest functions effectively within its context. This tripartite framework ensures a holistic approach to system development and deployment in the complex field of modern warfare.

System of Interest (SOI): The US military is a formidable institution, structured into six distinct service branches: the Army, Marine Corps, Navy, Air Force, Space Force, and Coast Guard. Each branch serves a unique purpose and operates in different domains. The Army, being the oldest and largest, is primarily responsible for land-based military operations. The Navy, with its impressive fleet, secures control of the seas and projects power across the globe. The Marine Corps, working closely with the Navy, specializes in amphibious operations and rapid response on land and at sea. The Air Force dominates the skies, providing air support and deploying forces swiftly. The Space Force, the newest branch, is dedicated to protecting US interests in space and maintaining space superiority. Lastly, the Coast Guard ensures maritime safety, security, and stewardship, operating under the Department of Homeland Security in peacetime and as part of the Navy during war. Together, these branches form a comprehensive military system, safeguarding the nation's security and interests both domestically and internationally.

Enabling System: The US military is enabled by a combination of factors that include advanced technology, a large and well-trained workforce, and a significant budget that allows for extensive research and development as well as technological advancements in areas such as precision-strike capabilities, intelligence, reconnaissance, and special operations. The military's strength is also reflected in its air power, with the US Air Force being the largest in the world, supported by the Army Aviation Branch. The US military's power is not just in its immediate force but also in its ability to sustain operations over time and adapt to emerging threats, making it a formidable global presence.

Context System: The military's capabilities are further enhanced by a global network of bases which facilitates rapid deployment and power projection. US military maintains a strategic advantage. Additionally, the USA is a member of international alliances, like NATO, which provide collaborative defense strategies.

Architectural Decisions Impacted by Mental Models

The structure of the US military is shaped by a myriad of decisions influenced by strategic, political, and economic factors. Strategic considerations include the need to counter global threats and maintain a balance of power, which can lead to adjustments in force size and capability. Economic factors play a crucial role as well, with budget constraints and funding levels directly impacting force readiness and modernization programs. The evolving nature of international relations, including shifts in alliances and partnerships, also necessitates changes in military posture and strategy. Cultural and societal values can influence recruitment, retention, and the integration of diverse service members, affecting the human capital that is essential to military effectiveness. Lessons learned from past conflicts and military engagements provide valuable insights that can lead to significant structural changes within the armed forces. These decisions are interconnected, each influencing the other and collectively shaping the dynamic and complex structure of the US military.

Sensitivity to Human Reach

The psychological and sociological dynamics of warfare, examining the interplay of morality, politics, and human nature, delve into the complexities of revolutionary movements, the justifications for terror, and the moral dilemmas faced by individuals in historical tumults.

4.3.6 *Sources of Complexity*

The complexity of modern warfare can often be attributed to emergent behavior, which refers to unexpected outcomes and patterns that arise from the interactions of simpler elements within a system. In the context of warfare, this means that even when individual soldiers and technologies function as intended, the collective behavior of a military force can lead to unpredictable and complex scenarios. This unpredictability is compounded by the rapid evolution of technology and the dynamic nature of geopolitical relations, making modern warfare a highly intricate and evolving challenge.

Modern warfare is often fought in increasingly urbanized battlefields, which are cluttered with civilians and civilian objects. This adds another layer of complexity as it requires careful planning and execution to minimize civilian casualties and damage to civilian infrastructure. The rise of state-like organized armed groups that leverage the governance vacuum created by failed or failing states is another factor contributing to the complexity of modern warfare. These groups often have access to advanced weaponry and employ unconventional tactics, making them a formidable adversary. The complexity of modern warfare is a result of the advent of hybrid warfare, the increasing number of variables on the battlefield, the role of politics and morality, the urbanized battlefield, and the rise of state-like organized armed groups. Understanding these complexities is crucial for any meaningful reform in the system.

4.4 Reflection on Chapter 4 Material

We discussed three man-made systems that require systems thinking to understand and influence: air traffic control, the US justice system, and a warfighting system. For each system, we tackled the complexity by identifying their relevant systems (context system, system of interest, and enabling systems) and highlighting their main: (1) elements, interconnects, and purpose; (2) influence of purpose, structure, and behavior on each other; (3) boundaries of the context system, system of interest (SoI), and enabling systems; (4) life cycle and evolution; (5) interaction and feedback within and between the relevant systems; (6) architectural decisions impacted by mental models; and (7) sensitivity to human reach. The reason we emphasized these features is to ensure that we have all the necessary basic terminology and understanding for learning the details of *systems thinking philosophy* in Part II that starts with *Systems Principles* in Chapter 5 next. Part II is all about how we can take advantage of *Systems Principles* to guide our thinking and behavior to understand real-world problems and the systems intended to operate in or deal with them in some way. This understanding can lead to innovative solutions and improvements in many areas of human endeavor.

Part II
Systems Thinking Philosophy

You think because you understand 'one' you must also understand 'two', because one and one make two. But you must also understand 'and'.—*Rumi*

Chapter 5
Systems Principles

Abstract Systems thinking is essential for approaching and creating sustainable solutions for real-world problems, as discussed in Part I of this book. The lack of systems thinking is evident almost everywhere where project costs and schedules are worse than planned or expected. Other common examples are where an anomaly in a small, seemingly uncritical part of the system causes a significant disaster and losses far beyond its cost or known significance. Chapter 3 gives easy-to-understand, convincing examples of real-world complex problems that require systems thinking. Unfortunately, our language, culture, and years of compartmentalized, disciplinary, and inside-the-box oriented education do not help! Encouraging systems thinking by pointing out the many examples of failures and fiascoes that could have been avoided is a good start but not enough. First, we must understand the details of clearly observable misconceptions or mental models that drive our linear compartmentalized thinking. These hamper our ability to find sustainable solutions to real-world problems. Then, we need reliable, practical, and actionable principles to guide our thinking, behavior, and methods for approaching various real-world problems. The idea is to apply principles that are always true to help maximize our understanding and improve the odds of success of our solutions. We call these *Systems Thinking Principles* or *Systems Principles* or *SPs*. We explain why we need the *SPs* in Section 5.1 followed by presenting each *SP* in detail in Sections 5.3 to 5.10. We recommend beginners to skip the rest of the chapter and move on to Chapters 6 and 7 that help develop an eye for the Systems Principles and how to apply them to understand real-world problems and find their best points of leverage.

5.1 The Need for *Systems Principles*

Our counterproductive linear compartmentalized mental models that reduce our chances of success as we approach complex problems are due to several factors, such as the architecture of our secondary and undergraduate education in addition to bad habits that are reinforced by the appearance of partial success and other external factors such as the culture that dominates in the environments we live and

K. Eftekhari Shahroudi et al., *Practical Systems Thinking*,
https://doi.org/10.1007/978-3-031-89394-0_5

work in. These misconceptions or bad habits that are common to all disciplines but particularly prevalent among STEM students are as follows:

Misconception: Divide and conquer always work. This is the misguided belief that the best way to improve a system is to divide it into parts, focus internally on the parts, and improve the whole by improving the parts. This mental model ignores that many real-world unwanted behaviors that we face today emerge from the interaction of the parts or the interaction of the SOI with other systems in the environment. An example of this misconception is a business leader who believes that significant improvement of the whole business comes only from a purely internal focus on improving the Marketing, Technology, and Program Management departments separately without worrying about how they do and should influence each other. Another example of this flawed mental model is for a music composer to focus internally on improving a single movement of several movements of a music score while ignoring the theme that threads or integrates the movements.

Misconception: Correlation is causation. This is a common misconception that the cause of some issues is near where the problem is observed (in time and space) and has the same pattern of behavior. For example, we would be wrong to conclude that we can boost air travel by boosting ice cream sales because both ice cream sales and air travel go up during the summer holidays! So, correlation is not the same as causation. Another example is only looking at airline pricing or airport system failures when air travel volumes drop. The cause may, for example, be a pandemic that is not directly inside the airline or airport systems but influences them in complex ways. Similarly, it would be misguided to credit a brand-new CEO of an airline company if profits rise within a week of the new CEO starting because the effects of the actions of the new CEO (and her new management team) are likely to show up after some time delay.

Misconception: Quick solutions give good results. Mastering tools and processes takes a lot of effort and focus, but it is counterproductive when they occupy all our efforts so that we lose sight of the problem's purpose or real nature. When faced with complex challenges, we pick a quick solution that easily fits our tools, processes, and methods and keep turning the crank, hoping the solution will fix the problem. So, the tools, processes, and methods occupy most of our available bandwidth and create a lock-in effect that reinforces further exclusion of real-world concerns. This takes time and effort away from understanding the complexities of the problem in the first place and considering a range of potential alternative solutions. For example, we have noticed that systems engineers may become overly focused on turning the crank on MBSE tools while forgetting that the primary purpose behind the systematic processes they follow (and the artifacts they create) is a better systemic understanding of the complex real-world challenge and solution.

Misconception: All models and their artifacts are correct and useful. The quote from George Box that "All models are wrong, but some are useful" teaches us to not always believe the quality and correctness of our modeling artifacts. An example of this mindset is using analytical tools such as finite element analysis (FEA) and computational fluid dynamics (CFD), which allow us to visualize the stresses or flow forces on parts in color graphics. The visually attractive results can be misleading because the analyst may think they are realistic and include all the required real-world considerations. Here, the analyst would be well served by continually trying to understand the boundaries of validating the results versus the real world. That way, any solutions based on the analysis will have a better chance of success in the real world.

Misconception: The scope of our thinking and perspective equals the scope of our deliverable. This is where we look at what we are responsible for delivering and assume that all our efforts and focus should be spent purely inside that scope. Even very experienced professionals can, unfortunately, exhibit this misconception because of the stress of delivering what they are responsible for. For example, imagine we are developing an emergency-use mobile lung ventilator that fits inside a carry-on-size briefcase to alleviate the shortage of available full-size ventilators in a COVID-19 pandemic. Suppose we limit our thinking and efforts to what is inside the SOI (i.e., the briefcase and its internal parts). Let us assume that we received a set of specifications from a pulmonologist in California that specifies the operating modes that describe how the ventilator controls the breathing of a COVID-19 patient. If, in this case, the internal design of the SOI occupies "all" of our thinking, attention, and efforts, then we will miss looking into whether pulmonologists in California use the same exact mode of operation or breathing treatment as pulmonologists on the East Coast. In this case, we will deliver an emergency ventilator that would potentially not be of much use across a large portion of the USA that does not have the same standard treatment for COVID-19 patients as in California. The more successful mental model is to think practically one or two levels outside of the area responsibility in addition to the SOI that we are responsible for delivering. In this way, we can predict and resolve the unwanted emergent issues that may arise from the interaction of the SOI with patients, operators, and other devices.

Misconception: Real-world problems are always well-defined. There are many real-world problems where the amount of information we have is not ideal. The information can be too much, too little, unreliable, or irrelevant to the problem. There are also many situations where there is not just one good answer, but we need to pick the best answer among several answers that all appear correct. However, our learning in school does not typically prepare us for these more complex situations, and we tend to use our personal biases to find a solution.

The above misconceptions and unhelpful mental models that many adopt are the unintended consequences of our compartmentalized education and work environments that reinforce pure internal disciplinary focus. Cross-cutting activities that encourage employing multiple disciplines and perspectives are not typically rewarded. We need to teach principles that are always true and useful regardless of the exact nature of the real-world challenges we face so that we may escape these unhelpful biases.

Correcting these misconceptions is traditionally through a significant investment of time and resources that involves learning and reflecting on material from an extensive list of systems thinking references given in Section 5.13. In addition, we need the opportunity to work on real-world problems to make the principles stick and become almost second nature. This level of investment and real-world practice opportunity is not usually available to many of us in higher-level education. The central idea behind the creation of this chapter and the *Systems Principles* in Section 5.2 is to jump-start the systems thinking learning process so that we can already use them to guide our thinking and approach as we try to understand real-world problems when we start our careers.

Next, we dive into the *SPs* first. Each principle has a statement, explanation, example highlighting its meaning, how it helps, and why we need it to guide our actions and behaviors. Section 5.11 describes the process behind developing the principles, and Section 5.12 outlines the requirements for each Systems Principle. Section 5.13 describes a list of excellent sources of information on systems thinking that we used as inputs to the development of the *Systems Principles* listed in Section 5.2.

5.2 *Systems Principles (SPs)*

A *principle* is a moral rule or a strong belief that (*should*) influences our thinking, behaviors, and actions. Here, we are mainly focused on thinking, behaviors, and actions that are helpful because they reliably improve our odds of success when we approach or face complex challenges and problems, regardless of their exact application domain or the combination of disciplines involved. To clarify, a *Systems Principle (SP)* is not a law such as Newton's first and second laws or first and second laws of thermodynamics, expressed in an equation or mathematical language that applies to every system within a defined range of conditions. Section 5.2 points to researchers who are progressing in this direction but have not yet reached the maturity required to derive the *SPs* directly. Instead, the *SPs* listed in this section come from the distillation of a large body of knowledge on systems thinking for which we have only real-world experiential or heuristic support, not scientific proof [18].

The list of *SPs* that we extracted from the large body of knowledge are as follows:

Note The words in brackets inside the *Systems Principles* are not strictly necessary but are added to help us understand the subtleties.

These *SPs* are the sources of truth that help us approach real-world complex problems. We need training on the depth and breadth of their meaning, implications, and applications. They form the primary basis for all the teachings in this book and, therefore, are invoked frequently throughout. In the following sections, we dive into each principle in detail. We explain the meaning of each principle and use simple real-world examples to highlight its subtleties. We then explain how each principle helps and why we need it to guide our actions and behaviors.

5.3 The *Holism Principle*

***Holism Principle*: (Complex) *systems* are (much) *more* than the *sum* of their *parts—elements*, *interconnects*, and *purpose*.**
This is the most universally acknowledged principle in the systems literature, and its roots go back to ancient philosophers who recognized the phenomena of *emergence*. This behavior arises from the interaction of elements in a system collaborating towards some purpose that is not directly predictable or understandable by the isolated knowledge of the parts: elements, interconnects, and purpose. Since the behavior is not a simple sum of the known internal characteristics of each part, to understand or improve a system, we need to understand the *emergence* that arises from their interplay. Hence, in a sense, the system is *more* than the sum of its parts. The greater this inequality, the more difficult the prediction of emergence.

Simple examples of the *Holism Principle*
A cake is created by measuring the ingredients in a certain proportion, mixing them in a particular manner, and baking them at a certain temperature for a specific duration. The individual or combined cake ingredients alone do not make a cake. It is how they combine to create a taste that emerges from the interconnection of the components to satisfy a purpose (see Fig. 5.1). In addition, if we remove a minor ingredient, it is possible to create a

(continued)

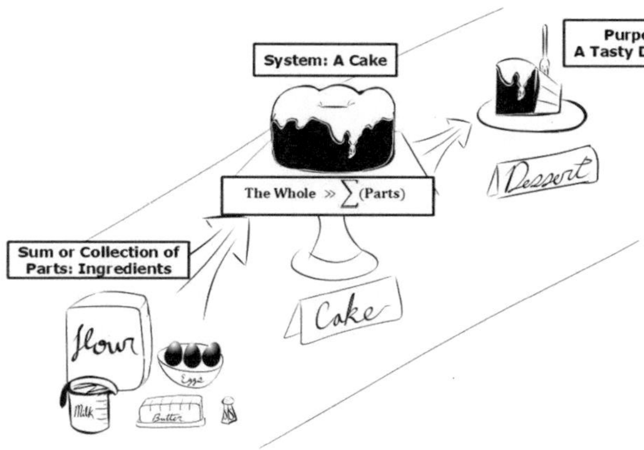

Fig. 5.1 *Holism Principle*: (complex) *systems* are (much) *more* than the *sum* of their *parts—elements*, *interconnects*, and *purpose*. Eating cake is much more satisfying than eating the components that go into it

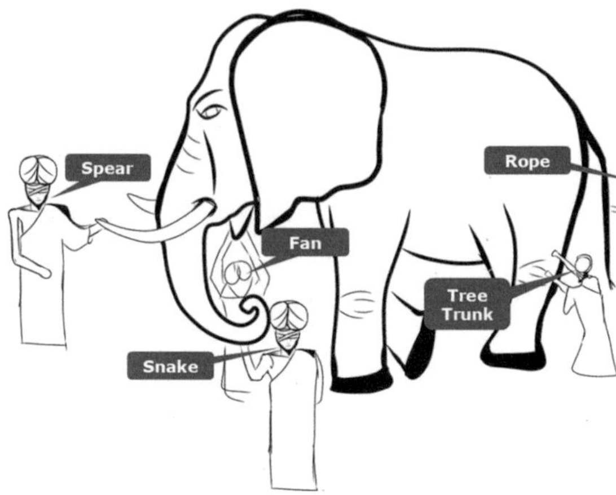

Fig. 5.2 The Sufi story of the *blind men and the elephant* highlights both the *Holism* and the *Architecture* principles

massive change in the cake's taste. For example, let's say that the chocolate component of the recipe weighs only 5% of the total weight of the cake. Linear or compartmentalized thinking would suggest that if we remove the dark chocolate (or replace it with white chocolate), we still end up with a cake that has 95% of the taste and satisfaction of the original cake since we only modified 5% of the cake! Another example of linear thinking is to imagine a baker competing with another baker who wants to double the satisfaction of their chocolate cake. If all they do is change the recipe to include better ingredients that are twice as expensive without changing the process and the way the ingredients are combined, they will end up with a cake that is only a little better but twice the cost!

The old Sufi story of the *blind men and the elephant* of Fig. 5.2 is another way to understand the *Holism Principle*: several blind men attempt to identify an elephant by touching the different parts of an elephant. They concluded snake, fan, spear, tree trunk, and rope based on touching the trunk, ear, tusk, leg, and tail, respectively. Their conclusions do not even come close to understanding the whole elephant: the leg (mistaken for a tree trunk) that is responsible for moving the entire elephant cannot even move anywhere, let alone move the elephant if it stands alone without interacting and collaborating with the other relevant parts of the elephant. So, we cannot just examine the individual parts of the elephant and extrapolate that understanding to understand the whole elephant without understanding how these parts are interconnected and how they work together. This is because the elephant is much more than the sum of its parts.

How does the *Holism Principle* help?
This principle encourages us to view real-world problems as systems, where emergence from the interaction of the parts is a strong possibility. It reminds us that subdividing and internally focusing on smaller parts are unlikely to be a winning strategy for understanding or predicting the emergent behavior because we are ignoring the *interconnection* between the *elements* and the *purpose*. We must be careful to consider the impact on the whole system and not inadvertently optimize a part to the detriment of the whole system. Our success odds improve when we balance our thinking and efforts by considering all three.

Why do we need the *Holism Principle* to guide our thinking and actions?
Our dominant approach to solving problems today is to subdivide the problem into smaller parts and apply focus and resources inside those parts to solve the whole. This approach has worked well in the past on more straightforward issues. That is how compartmentalized linear thinking became part of our problem-solving culture today. Unfortunately, even predictable matters can become complex and show strong emergence to cause more problems! We can use this principle to avoid falling into this linear compartmentalized thinking trap.

5.4 The *Architecture Principle*

***Architecture Principle*: The *purpose*, *structure*, and *behavior* of systems *influence* each other.**
When we are faced with a real-world complex problem (and the systems that deal with or operate in them), we need a strategy to find the best points of leverage. Otherwise, our understanding will likely lead to solutions that are impractically expensive, time-consuming, or counterproductive, so they make the situation worse or cause new problems. The Architecture Principle tells us that since *purpose*, *structure*, and *behavior* of systems *influence* each other, then interfering with one of these will not be self-contained and will ripple through to the other two. Finding the best leverage points then requires awareness and understanding of this inter-influence.

Simple examples of the *Architecture Principle*

The simple pieces in Fig. 5.3 could be nailed or connected together as a chair, a ladder, or various other forms. However, not every structure has a suitable function for any given purpose. For example, a spiral form with guard rails on the left is preferable to a super tall straight ladder for children to climb a very tall tower safely. The straight ladder to the right is convenient for a contractor because it can be carried on a truck. The multipurpose ladder on the top right can satisfy multiple purposes because its design allows it to change its form (or structure) even if the basic elements remain the same. Yet none of these are suitable for safely and efficiently servicing tall street lights: the truck-mounted actuated cabin on the right. Identifying the purpose first is very helpful for determining the best structure, i.e., the interconnections required between the elements. Changing the purpose changes the needed structure and the resulting way it functions or behaves. So, Fig. 5.3 shows how the different *purposes*, *structures*, and *behaviors* influence each other for a variety of applications.

We can also use the Sufi story of Fig. 5.2 to understand the *Architecture Principle*: the external or touchable parts of the elephant, by themselves, are unlikely to give easy hints of the *pattern of behavior* of the elephant or its *purpose*. Major systems thinking texts (Dana Meadows [12], Peter Senge [17], Sterman [20]) teach that the purpose drives (or determines or causes) the structure that drives (or determines or causes) the pattern of behavior. Therefore, the higher up this chain we focus our efforts, the higher the leverage to improve the system. In the cake example, this implies that changing the recipe and ingredients (structure) would have a more significant effect or higher leverage on the taste and satisfaction than just changing the way the cake looks and tastes with food coloring and flavoring. However, these excellent references also warn us that higher leverage locations are not

(continued)

Fig. 5.3 *Architecture Principle*: the *purpose*, *structure*, and *behavior* of systems *influence* each other. The structure and behavior (or function) of the ladder depend on its purpose and vice versa! Even the multipurpose ladder has to change its structure (shape) and behavior (how it functions) to meet different purposes

typically obvious just by looking at the behavior. This is similar to the part about touching an elephant in the Sufi story, where it is hard for the blind men to deduce the pattern of behavior of the whole elephant just by touching the parts, let alone finding the purpose of the entire elephant.

In fact, to an external observer, the true *purpose* of a system is typically the hardest to understand because one would typically have to "learn" the *patterns of behavior* and the *structure* that cause it and are influenced by it. Imagine we are a competing baker but not the creator of the cake shown in Fig. 5.1 yet motivated to figure out why it tastes so much better than our cake. We do not know the other creator's recipe. Let's also assume that the taste did not directly reveal the ingredients. The *Architecture Principle* implies that our leverage is higher or our chances improve by understanding the cake's recipe (*structure*) instead of trying to directly change the taste of our cake to match the original (e.g., by sprinkling flavor powders). Although the original cake's recipe (and what the original creator was thinking) is not available to us, we may learn the recipe by combining our baking knowledge with exploration and experimentation of various recipes until we match or surpass the taste and satisfaction of the competing cake. This is called "learning," covered by the *Learning Principle* later in this chapter.

How does the *Architecture Principle* help?

The *Architecture Principle* guides us to not just make a change in one of the features of the architecture without intentionally considering how this impacts the other two at the same time. If we do not like how a system behaves, then the best place to interfere with the system may not be to directly change the unwanted *behavior*. Changing the *purpose* or *structure* that influences the *behavior* may be much more practical, convenient, or elegant.

Why do we need the *Architecture Principle* to guide our thinking and actions?

We often tend to pick and implement quick solutions. The *Architecture Principle* reminds us that there is more than one place to look for the best place to interfere with systems: *purpose*, *structure*, and *behavior*. It also reminds us that a change in any of the three features of the principle can ripple through to the other two. So the *Architecture Principle* also helps us avoid making the problem worse elsewhere when we implement a narrow solution.

5.5 The *3-Systems Principle*

***3-Systems Principle*: *Systems* come in (at least) *three*: *context systems*, *systems of interest (SoI)*, and *enabling systems*.**
The 3-Systems Principle is the reason why we have an "s" in system*s* thinking versus the singular form system thinking. Recall that we call the system we are trying to understand or deliver the *system of interest* or *SoI*. It turns out that the success, failure, performance, reliability, profitability, and any other significant property or characteristic of the *SoI* do not depend entirely on what is inside the boundaries of the *SoI*. It also depends on the interaction of the *SoI* with at least two other systems: the *context system* and the *enabling system*. The *context system* is the system that the *SoI* fits into or the environment that the *SoI* operates in. The *enabling system* is a system that is outside the *SoI* but interacts with or supports the *SoI* by exchanging material, energy, signals, and information with it.

Simple examples of the *3-Systems Principle*
For a heart specialist who is responsible for troubleshooting issues with the proper functioning of the heart such as tachycardia, their *system of interest* (SOI) is the heart. In Fig. 5.4, the body is the system into which the heart or the SOI fits. So the body is the *context system*. The proper functioning of the heart (SOI) is not only a function of its internal components but also depends on interactions and exchanges with *enabling systems* such as the circulatory, lymphatic, pulmonary, and gastric systems. So, we must include sleep, diet, and exercise among the factors considered for troubleshooting the heart, not just the parts and factors that are within the heart.

Tachycardia is when the heart races to abnormally high rates, say, above 160 beats per minute, that in some severe cases can keep going for a couple of weeks without intervention. Tachycardia can cause severe lightheadedness, fainting, or dropping to the floor because at very high pumping frequency the heart cannot pump oxygenated blood to the brain. If the high frequency continues for a while, it makes the person feel tired as if they have climbed a steep mountain. Some heart specialists who are only focused purely on the interior boundary of the heart (*SOI*) may suggest an expensive and intrusive heart surgery to burn some nerve endings inside the heart (e.g., by ablation) to hinder the nerve pathways inside the heart that are involved in the patient's tachycardia episodes. In the operating room, by applying electric signals to the heart, they encourage the heart to go into a tachycardia mode of behavior.

(continued)

Then they start zapping the nerve endings at the top and bottom of the heart until the heart returns back to normal rhythm and can no longer be encouraged to enter the tachycardia mode. The success rate of this procedure is high, but unwanted side effects are possible including the rare possibility of death during heart surgery. This is an example of linear cause-and-effect thinking: i.e., the cause must be close to where the problem shows up inside the SOI, both in terms of location and time. However, the true cause or the best solution may be in an entirely different location outside the heart (SOI) or at an earlier time (or both), inside the *context system* or *enabling systems*. For example, some patients discovered that by controlling their caffeine and alcohol intake,

(continued)

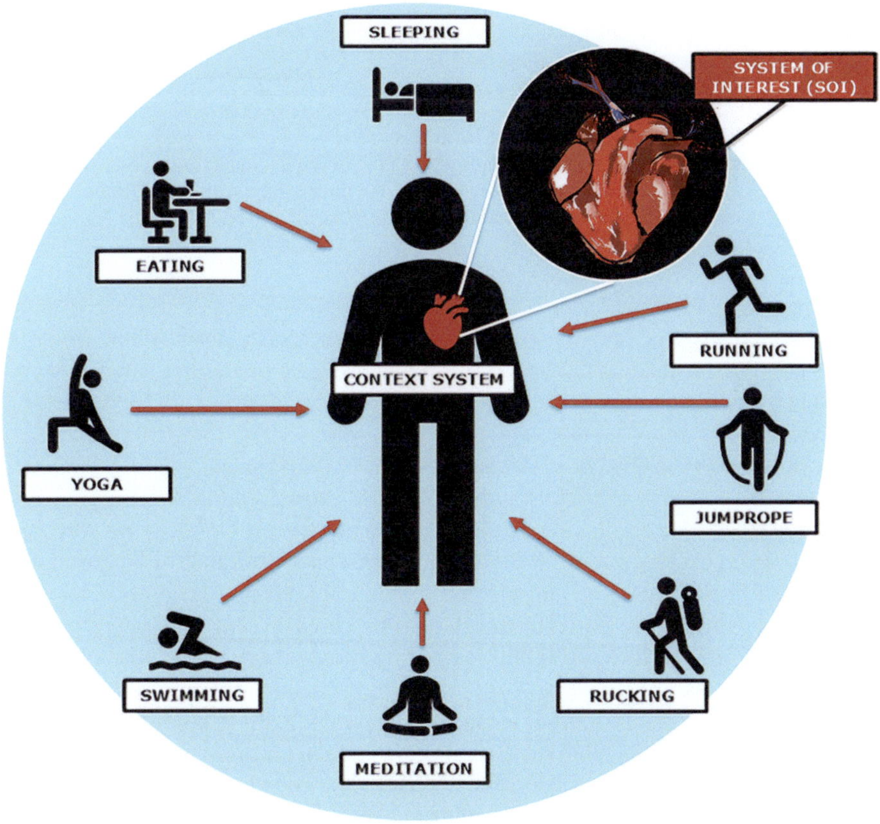

Fig. 5.4 *3-Systems Principle*: *systems* come in (at least) *three—context systems*, *systems of interest (SoI)*, and *enabling systems*—the three systems where the SoI is the human heart. Heart problems are not always caused by or best resolved by only treating parts of the heart or going to the heart surgeon! Eating, sleeping, and exercising are good candidates to add to our considerations when we troubleshoot heart problems

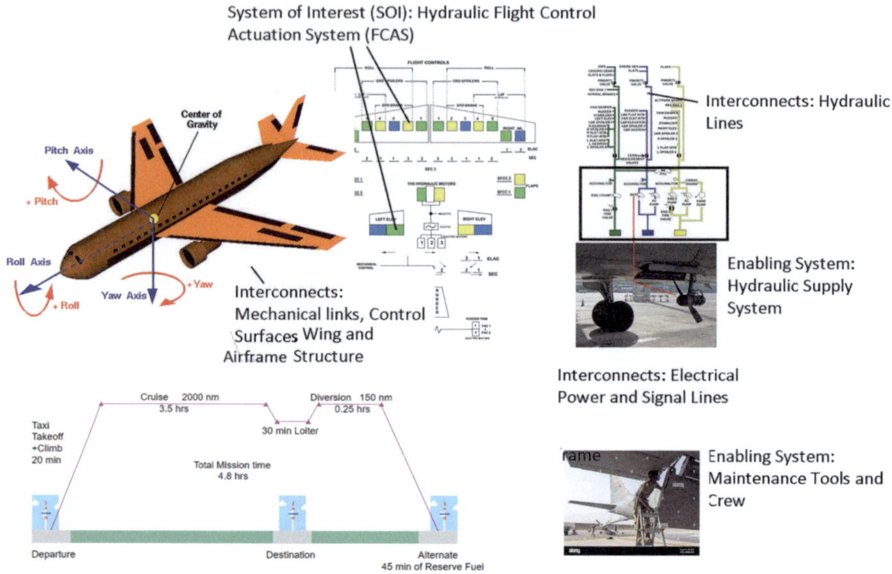

Fig. 5.5 *3-Systems Principle*: *systems* come in (at least) *three—context systems*, *systems of interest (SoI)*, and *enabling systems*—the three systems where the SoI is a hydraulic flight control actuation system (FCAS). The FCAS supplier has to understand the interactions with the airframe (context) and enabling systems (hydraulic supply, maintenance systems, etc.) to be successful

their tachycardia symptoms severely improved or totally disappeared which was the case for the lead author of this book. Caffeine or alcohol entering the gastric system (enabling system) would not normally be a concern for a linear-thinking heart surgeon whose actions and interests are focused primarily inside the heart (SOI). A systems thinking heart specialist would however balance their efforts and focus and would be interested in looking for causes and solutions located outside of the heart in case the problem can be solved by a simple change or controlling of some input outside of the cardiovascular system. Limiting caffeine and alcohol intake may not be the solution to all tachycardia cases, but patients who would be cured by this approach would certainly prefer it to a $100,000 surgery that can cause other side effects including death!

Figure 5.5 shows a different example of a hydraulic flight control actuation system (FCAS) responsible for the roll, yaw, and pitch motion of an aircraft. The FCAS is one of the most critical systems on board because the aircraft needs to still be able to maneuver and land even when all other major systems (such as engines) have failed. To reach super high levels of reliability, the FCAS has multiple points of redundancy so that it can still function in the

(continued)

event of an actuation line or a hydraulic line failing. For example, we can see in the middle-top picture that the right and left aileron surfaces on the wing that control roll use two actuators of different colors. Here, the color refers to which of the three redundant hydraulic supply systems the actuator is connected to so that, for example, the blue line loses hydraulic pressure, then the actuator connected to the yellow line will still be able to function and the aircraft will not lose roll control. Now, suppose a supplier is responsible for delivering the aileron actuators which is their SOI. If they only care about their SOI and focus their understanding purely on the internal workings of the SOI they are responsible for, then they can miss complex interactions between their SOI and all the other systems on aircraft their SOI interacts with which can lead to failure. The *3-Systems Principle* guides the FCAS supplier to attempt to understand one or two levels outside their SOI to ensure that their SOI is successful. For example, in this case, the *3-Systems Principle* urges the FCAS supplier to understand the relevant interactions with mechanical interconnects and components, electrical interconnects and components, hydraulic interconnects and components, and the actuation systems for the rudder and elevators that the FCAS supplier is not supplying.

How does the *3-Systems Principle* help?

This principle ensures that we try to understand the *three systems* in the minimum and not just spend our efforts and focus inside the *SoI*. In other words, this principle forces us to think one or two levels outside of the *SoI* that we are responsible for. Note that the *3-Systems Principle* is emphasizing *at least* three systems. In practice, we may need to consider many more systems. For example, we may have multiple *enabling systems*. We can also bring together a set of systems for a task that none of the systems can accomplish on its own. Here, each constituent system keeps its own management, goals, and resources while coordinating with the other systems. In this case, we will have multiple *SoI*s, *enabling systems*, and *context systems* to manage to ensure that each system works in harmony with the others. So depending on the complexity of the problem we face, we may need to consider more than three systems!

Why do we need the *3-Systems Principle* to guide our thinking and actions?

We tend to only worry about and focus on what we are responsible for or delivering. In practice, this behavior has led to many failures or fiascoes, either immediately or over time, that we could not predict or avoid because of our

(continued)

limited focus on the *SoI*. This principle expands our understanding beyond the *SoI* and forces us to identify and understand all the relevant systems outside the *SoI* that significantly influence its success or failure.

5.6 The *Boundaries Principle*

***Boundaries Principle*: *Systems boundaries* depend on *perspectives*, *purposes*, and *roles*.**
Einstein emphasized that clearly defining a problem is a critical step to solving it: "If I had an hour to solve a problem, I'd spend 55 minutes thinking about the problem and 5 minutes thinking about solutions". By analogy, clearly defining the boundaries of the *three systems* significantly impacts how the problem is defined and, therefore, very influential to our understanding of the problem and our ability to solve it. The *3-systems* boundaries identify precisely what is included inside each of the three systems. This principle tells us that the best systems boundaries or ways to frame the problem depend on the individual's perspective, purpose, and role.

Simple examples of the *Boundaries Principle*
If we are driving 70 miles from downtown Denver to Fort Collins after work around 5 pm, typically, the most useful initial boundary for the system of interest before starting is something like the map on the left side of Fig. 5.6 with key information overlaid such as air quality, location of gas stations, time to destination, etc. However, as we start our journey and continue to shift our position, we may also typically want to shift the boundaries to move with the car. We all understand that a good navigation app allows zooming in/out to fit our current location, our purpose, and our responsibility to arrive safely at the destination.

Different roles may disagree on the best boundaries of the three systems because their area of responsibility is different, even if they are working for the same enterprise or for the same city as shown on the right of Fig. 5.6. For example, a builder may think of a stretch of open road as what he is trying to build, deliver, and get paid for. However, the Department of Transportation of the City of Denver, which has the responsibility to ensure safe and efficient travel for the city, will be better served with a broader view showing how heavy traffic jams emerge from the interplay of major highways, events, public/private transportation systems, and events like hail and accidents.

(continued)

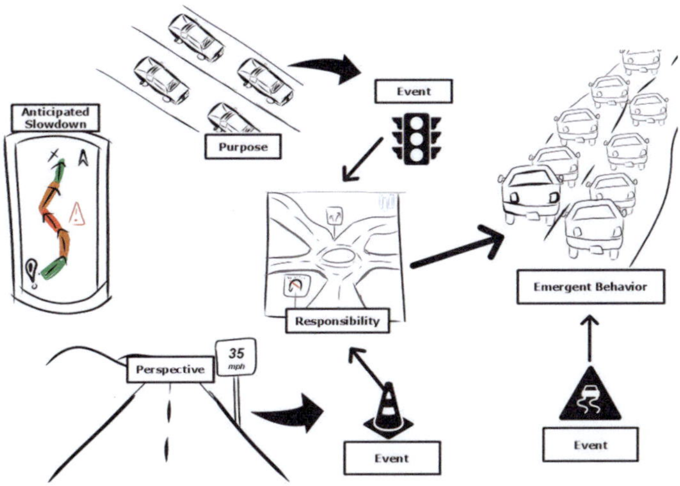

Fig. 5.6 *Boundaries Principle*: *systems boundaries* depend on *perspectives*, *purposes*, and *roles*. Highway transportation. A driver, highway builder, and city planner may all be concerned about heavy traffic. Still, their best way to view the problem is different because their perspectives, purposes, and roles differ

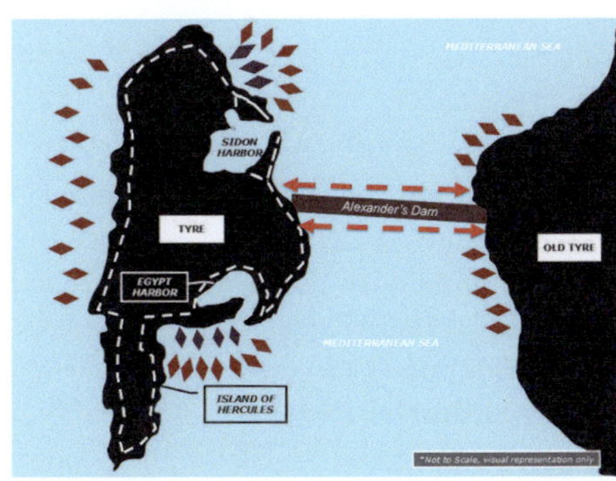

Fig. 5.7 *Boundaries Principle*: *systems boundaries* depend on *perspectives*, *purposes*, and *roles*. Understanding where to draw the systems boundaries helped Alexander the Great find the best leverage to defeat the Persian Navy on land, not at sea, where the Persian Navy had a big advantage

Figure 5.7 visualizes the battle of Tyre that occurred in 332 BC for yet another example of the *Boundaries Principle*. At this time, the conventional fighting strategy was that naval fleets would face each other head on. Normally, the bigger and more advanced navy would win the battle. In this case, the Persian fleet was superior to the fleet that Alexander commanded. Yet Alexander defeated the Persian Navy by capturing Tyre, the island port that supplied the Persian fleet, versus fighting the larger more advanced fleet head on. Alexander did not consider the entire enemy as one large obstacle. He drew the boundaries that separated the fleet (SOI) from the supply port

(continued)

(enabling system) in such a way that enabled identifying an easier target or the highest point of leverage that would result in the collapse of the enemy fleet. Alexander framed the boundaries of the problem in a way that a more elegant highly leveraged strategy became more evident: disable the enemy supply base (*enabling system*), not the fleet the (*SOI*), and the whole system would collapse! Had the Persian Navy known about this principle, they would have realized that the enemy with a fleet disadvantage who has a different goal and perceptive might frame the problem differently and not fight a traditional battle of numbers and capability.

How does the *Boundaries Principle* help?
This principle teaches us that there is not just one optimal way to define the systems boundaries for everyone involved. The most useful boundaries are a function of our perspectives, roles, and responsibilities. It helps us view the boundaries of the systems from different perspectives and be willing to adjust our definitions accordingly. This principle guides us to find the most beneficial framing of the complex problem as systems.

Why do we need the *Boundaries Principle* to guide our thinking and actions?
As individuals or teams, we tend to be rigid about the way in which we frame or view a real-world problem as a system. We also tend to assume that this best framing of the problem is the same for everyone else involved. We need the *Boundaries Principle* to encourage us to be flexible with our framing and also be sensitive to the framing of other stakeholders.

5.7 The *Evolution Principle*

***Evolution Principle*: Systems have *life cycles* and they *evolve*.**
All real-world problems and the systems that connect, interact, and operate in them in some way are not static in their purpose, structure, and behavior. The two most common patterns of change are *life cycles* and *evolution*. We need to recognize and anticipate the consequences of these common patterns of change for the specific problem at hand.

Simple examples of the *Evolution Principle*
Understanding the life cycle of a frog in Fig. 5.8 involves identifying and studying each stage of its development, the behavior of each stage, and how it transitions between those stages. Understanding the frog's evolution involves observation and recording of structural and behavioral changes of these stages over many generations, including how the various forms interact with other species and the ecosystem.

For example, the bat sonar capability shown in Fig. 5.9 has evolved to become a very accurate tool for hunting flying insects in the dark. Yet, this evolution has caused a counterbalancing evolution in the tiger moth which has evolved the capability of jamming the bat sonar. The picture on the right of Fig. 5.8 is from a test video where a tiger moth avoided being captured at the last moment by creating loud clicks that jammed the bat sonar system.

Similar to natural systems, it is very helpful to realize that man-made systems also have life cycles and evolve. Consider a football as a system in Fig. 5.10. The life cycle of a football "system" includes acquiring the various materials (tangible elements), processing these materials, and integrating them to make the specific prolate sphere shape, using and maintaining (brushing and conditioning) the football, and then finally disposing of the football. The evolution of the football itself and the game of football are heavily intertwined. When the use of a forward pass in the game was

(continued)

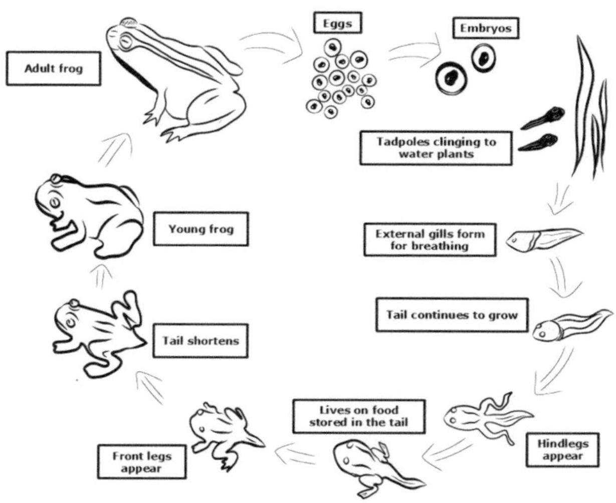

Fig. 5.8 *Evolution Principle*: systems have *life cycles* and they *evolve*. Awareness of the systems' *life cycles* and their *evolution* enables sustainable solutions

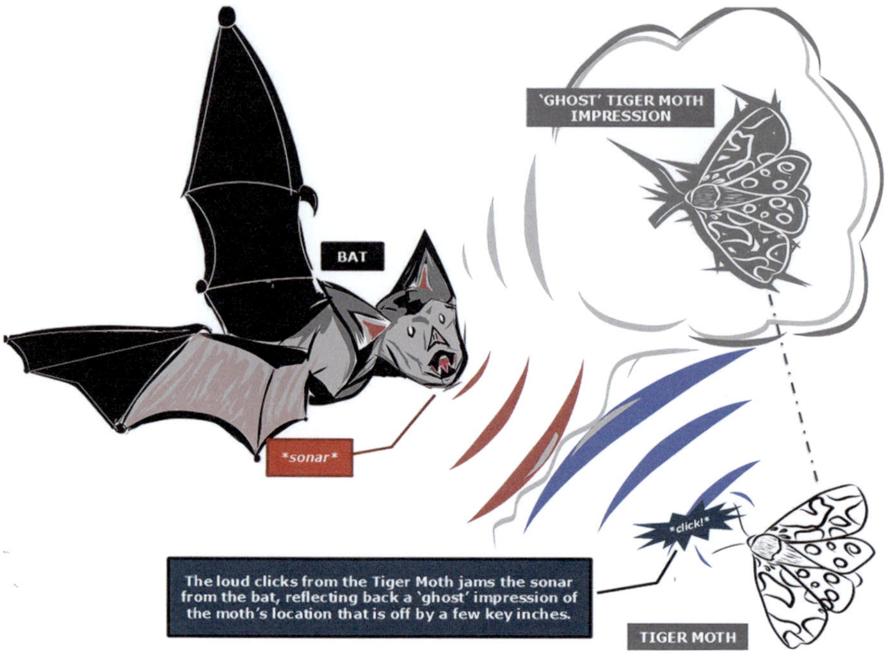

Fig. 5.9 *Evolution Principle*: systems have *life cycles* and they *evolve*. The bat sonar evolved to detect flying insects. But then the tiger moth evolved to jam the sonar to mislead the bat. So, the evolution of bat and moth capabilities influenced each other

approved, the shape of the ball needed to evolve to meet the new customer's need for improved aerodynamics.

The concepts of *adaptability*, *modularity*, and *resilience* are very relevant to designing the architecture and planning the future operation of man-made systems because these are the ways in which we can anticipate and be ready for potential future changes. For example, some cars incorporate software to adapt the way it moves to the driver's driving style. Tesla EVs allow remote software updates to regularly add new features or change the FSD software to as better self-driving modules become available. Another good example is how the International Space Station used a *modular* design to plan for future expansion with new modules and technologies as they became available decades later. We mention these concepts here because this principle, in particular, has strongly impacted the systematic architecture and design of engineering solutions.

(continued)

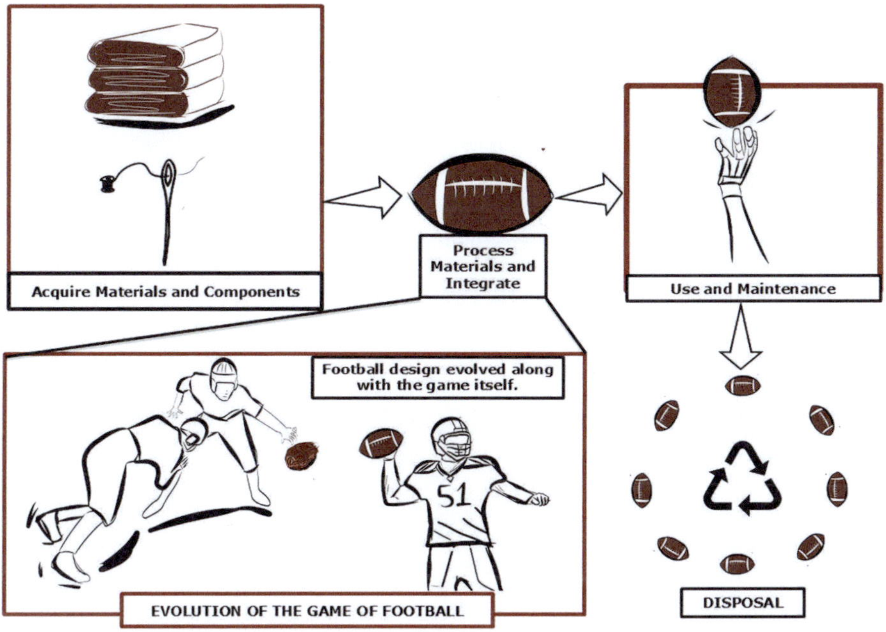

Fig. 5.10 *Evolution Principle*: systems have *life cycles* and they *evolve*. The game of football evolves and the football itself has a life cycle

We explained early on in the preface that here we are focusing more on the philosophy, approach, and rigorous methods to understand real-world problems and systems rather than create, architect, and design systems. Thinking about problems as systems and creating systems to solve them are two sides of the same coin and strongly depend on each other. Systematically creating solutions, however, requires more in-depth domain-specific knowledge and transdisciplinary frameworks, languages, and tools that are, for example, found in systems engineering. Teaching systems engineering is beyond the scope of this book. There are excellent references like those by Azad Madni [10, 11] and the *INCOSE Systems Engineering Handbook* [6] and the Model-Based System Architecture Process (MBSAP) by Borky and Bradley [2]. These focus more on the best ways and processes to systematically create, develop, implement, operate, and eventually retire systems intended to solve real-world problems.

How does the *Evolution Principle* help?
Understanding the life cycle and evolution of the *three systems* helps us anticipate the wanted and unwanted impact of our solutions throughout the life cycle stages and over time. This helps us predict what might happen next, prepare for upgrades when needed, and use our resources wisely. Knowing this principle drives proactive planning for risks that may come up and a dedication to continuous improvement to ensure our understanding and our solutions remain useful.

Why do we need the *Evolution Principle* to guide our thinking and actions?
Updating our mental models takes a lot of effort and is difficult. Hence, we would naturally tend to think of problems as static targets versus challenges where purpose, structure, and behavior are changing in stages and evolving. We need this principle to force us to put in the effort to understand the changes to improve our chances of predicting the wanted and unwanted consequences of our solutions over time.

5.8 The *Emergence Principle*

***Emergence Principle*: *Interactions*(including feedback) within and between the three systems cause (sometimes unwanted) *emergent behaviors*.**
Emergence is the phenomenon where the properties, characteristics, or behavior of the whole are not directly or easily predictable from the knowledge of the parts alone. It comes from the interconnection, interaction, or exchange of material, energy, and information between the parts of a system or between the three systems. Emergence is often caused by feedback loops that cause reinforcing and balancing behaviors.

Simple examples of the *Emergence Principle*
Our world and life on this planet rely on the emergent behavior formed around the bond of hydrogen and oxygen that occurs here on Earth. Water and oxygen are essential to supporting this complex world. The very creation of this is an

(continued)

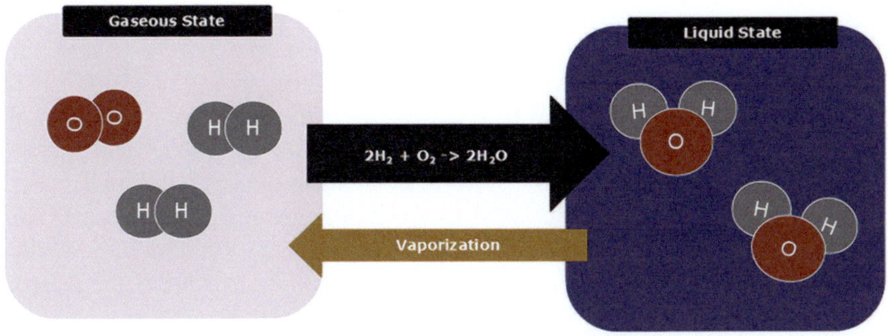

Select Properties	Hydrogen	Oxygen	Water
Boiling Point	-253	-183	100
State at Room Temperature	Gas	Gas	Liquid
Flammability	Explosive	Necessary for combustion	Used to extinguish flame

Fig. 5.11 *Emergence Principle*: *interactions* (including feedback) within and between the three systems cause (sometimes unwanted) *emergent behaviors*. What emerges from the interaction of hydrogen with oxygen molecules (water) bears no similarity with them

example of emergent behavior. Alone, hydrogen and oxygen molecules exist in isolation and have their own properties. Yet, when combined, they no longer behave as individual molecules but instead as a new emergent substance, water. Hydrogen is a gas that is lighter than air. When it reacts with another gas oxygen, the result is water that has very different properties to its components oxygen and hydrogen. Figure 5.11 compares the boiling point, state at room temperature, and flammability of hydrogen, oxygen, and water. Hydrogen is explosive, oxygen helps to burn, but water is a flame extinguisher! Clearly, what emerges from the reaction of oxygen and hydrogen is not linearly or explainable by summing their properties.

Another example is the story of the Borneo cats. During the 1950s, the island of Borneo, Malaysia, was battling a serious outbreak of malaria which is a mosquito-borne disease. They asked for help from the World Health Organization who responded with a tried-and-tested solution that was to spray DDT which was known to be effective against mosquitos. They sprayed DDT in a thick layer where the mosquitos lived which was on the interior walls of their thatched-roof homes. As expected, the DDT was effective in killing the mosquitos, and malaria problem was greatly reduced. The WHO was

(continued)

congratulated on once again solving problems. However, as time passed, the thatched roofs of the homes sprayed with DDT began to fall apart. Upon further investigation by the WHO, it was discovered that the population of a thatch-eating caterpillar, who were able to avoid the DDT-laced thatch, had greatly increased, but the population of their parasites, a small wasp, who did not avoid the DDT laced thatch had decreased. Again, the WHO decided a simple solution to the problem which was to use a more powerful poison that would kill the mosquitos, the caterpillars, as well as the wasps. This again was effective in the short term. However, the long-term effects showed up in other parts of the food chain. Since DDT stays in the body for years, the wasps that didn't immediately die from the DDT were eaten by the local geckos. Since the geckos were resistant to the effects of DDT, it then accumulated the DDT in their body. Their predators, the cats, ate the geckos and also rubbed against the walls sprayed with the DDT and licked their fur to clean themselves, which cumulatively caused them to die. When the cats died, their other prey, the rat, population increased dramatically. The rats ate the crops which decimated the food stores for the humans who now starved. This led to rat poison and continued air-dropped replacement cats to solve the problem. This story highlights how using tried-and-true linear solutions doesn't always lead to a long-term solution and can create new and worse problems. This could have been solved faster if the environment and the ecosystems that existed were taken into consideration. Figure 5.12 shows the complex emergence caused by the two unwanted feedback paths.

Yet another example of the *Emergence Principle* is shown in the cartoon in Fig. 5.13 (first used by John Sterman [20] in his system dynamics lectures at MIT). The cartoon shows a manager who is only willing to think and look inside the red box of his immediate vicinity, i.e., he has a limited mental model in that he is blissfully unaware of interactions and feedback between the systems that can lead to major unwanted behavior. He is feeling pressure, and to create more space, he pushes one of his problems out of his area of responsibility. Unbeknownst to him, this interacts with the other elements in the system context and comes back through a feedback loop with delay to hurt him at a later time.

The manager is initially happy with his progress in creating more space and is not worried or focused on the consequences of this action outside the area of his immediate responsibility which is the red box. However, his action puts a chain of events in motion outside his immediate vicinity and eventually comes back to hit him on the head!

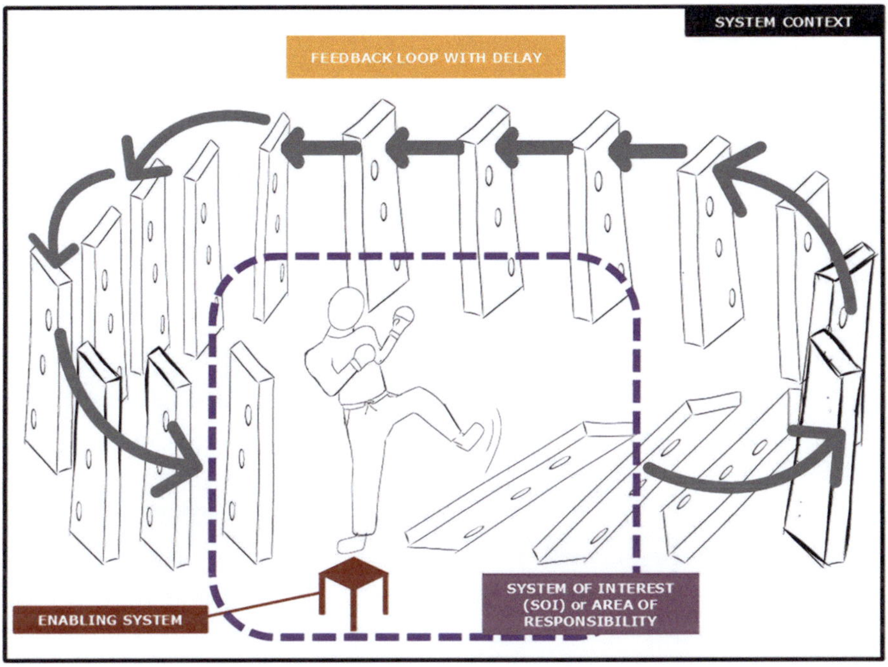

Fig. 5.12 *Emergence Principle*: *interactions*(including feedback) within and between the three systems cause (sometimes unwanted) *emergent behaviors*. This is an example of unexpected and unwanted *emergence* due to delayed feedback of the manager's actions, where the manager's mental model is limited to only the *SoI*: the area he is responsible for

How does the *Emergence Principle* help?

The *Emergence Principle* tells us that if we have a limited mental model or imagination of a complex problem, issue, situation, or system, we will fail to predict the short- and long-term consequences of our decisions or actions that can come back to harm us at a different time or location of where the decision or action originated. We call this effect feedback, and it turns out many wanted or unwanted emergent behaviors over time are driven by feedback as shown in this example. Understanding this principle helps us try to identify and predict the feedback loops that our actions or decisions will put into motion in the hope of improving the desired behavior, predicting unwanted consequences, and avoiding them.

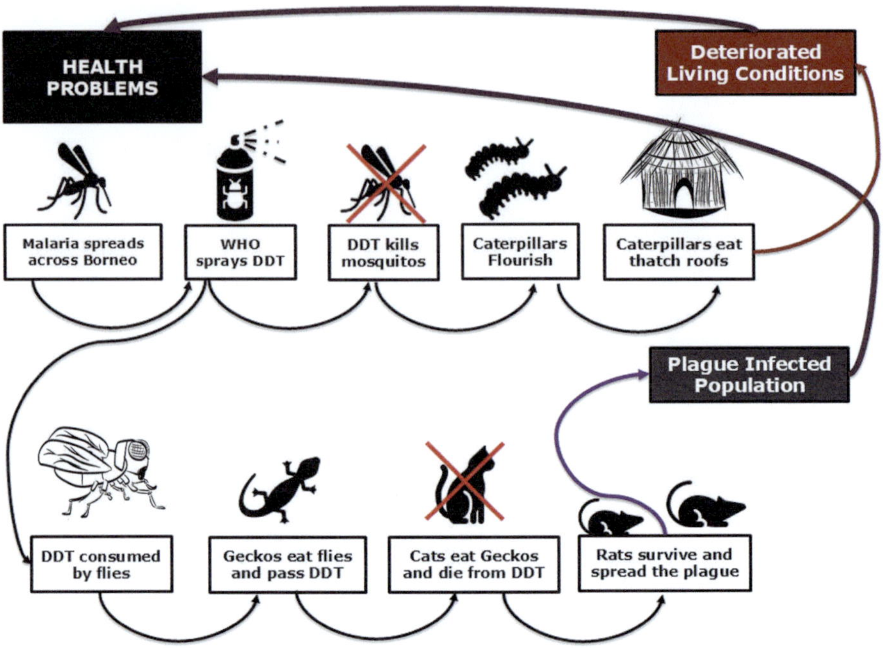

Fig. 5.13 *Emergence Principle*: *interactions* (including feedback) within and between the three systems cause (sometimes unwanted) *emergent behaviors*. Real-world example of a well-intended policy to reduce malaria causing the emergence of new health problems over time

Why do we need the *Emergence Principle* to guide our thinking and actions?

The term "root cause" comes from a linear-thinking mindset where multiple unidirectional chains of cause and effect merge to contribute to a problem we are trying to understand. The fishbone diagram of Fig. 1.2 is a good example of this mental model where the idea is to find the most significant factor among all the factors on the fishbone diagram. This linear mindset unfortunately comes short in the face of most real-world problems because a common cause of emergence is one or more feedback loops that interact to create the emerging behavior. The *Emergence Principle* forces us to always identify the relevant feedback loops to gain a richer more realistic understanding than linear thinking allows.

5.9 The *Learning Principle*

***Learning Principle*: (Common) *Mental models* influencing human *decisions* about *architectures* are *learnable* and *adaptable*.**
We have noticed that understanding, defining, or identifying the right real-world problem to pursue is a very reliable predictor of project success. But where does such understanding come from? The answer is that it is created by systems thinking *mental models* that strive to view and deal with real-world problems as systems. This means that systems thinkers attempt to *learn* how the *architecture* of systems are interconnected with the real world to create the part of the problem or system that we can see or measure. The *Learning Principle* assures us that this effort is not futile and that our mental models can adapt or learn until they reach a fruitful state where they can aid in understanding, defining, or identifying real-world problems.

Simple examples of the *Learning Principle*
What would the blind Sufis in Fig. 5.2 have to do to be still able to identify the elephant, assuming we cannot cure their blindness? The first obvious suggestion is for them to understand that they need not solve the problem alone and to communicate their understanding, speculations, and learnings collaboratively. Next, they would have to understand the patterns of behavior somehow. For example, keep touching (more parts), listening, smelling, and moving with the elephant throughout the day. At some point, a *pattern of behavior* will emerge: the legs (tree trunks) move together in a particular pattern, and so do the head (sway) and the tail (wag) side to side with the forward motion. Soon, they will conclude that this is a huge and strong animal. If at least one Sufi has known or heard about an elephant before, then the various motion and sound patterns might make the puzzle easier. If none of them had ever known about an elephant, they would proceed to map the parts better and how they are interconnected. They would establish how these parts cause the patterns of behavior identified earlier and then move on to finding the purpose. Now, how far should the Sufis go to try to solve this problem? Well, depending on the task, they can keep going until a practical or justifiable level of effort has been spent. But no matter when they stop, trying to find the patterns of behavior and relate them to a structure and purpose will have yielded much better answers than their original isolated touch-and-conclude approach!

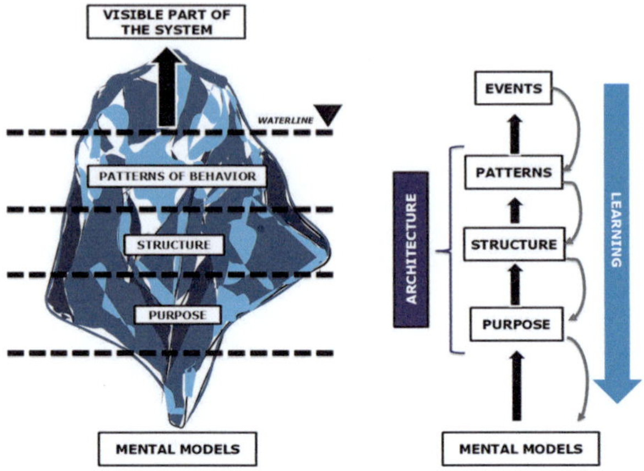

Fig. 5.14 *Learning Principle*: (common) *mental models* influencing human *decisions* about *architectures* are *learnable* and *adaptable*. Similar to the iceberg, the visible or measurable part of real-world complex problems (events) does not offer sufficient understanding or opportunities for improvements. Systems thinking encourages us to put effort into diving deeper to learn the patterns of behavior, the structure that causes those patterns, and the purpose that created the structure. Ultimately, the highest leverage is achieved by learned and adapted mental models that define the purpose

The left side of Fig. 5.14 shows an iceberg where only a small part is visible. There is a much more significant part below the surface that causes the part above the surface to rise so we can see part of what is floating. By analogy with the iceberg, the understanding required to approach and solve complex challenges is not directly knowable from the measurable event or data. We need to dive deeper to identify the *patterns of behavior* that cause or explain the visible or measurable information. However, as we saw in the *Architecture Principle*, the pattern of behavior is itself influenced by the *structure* and the *purpose*. Systems thinking teaches us to not always react to the measurable and more easily available information directly and quickly. Instead, it guides us to balance our efforts for a better understanding of the architecture of the problems: its *purpose*, *structure*, and *patterns of behavior*.

The deeper we dive or the better we understand the *architecture*, the stronger our leverage on changing or improving the part that we can directly measure. The right side of Fig. 5.14 shows this learning but goes deeper than just the *architecture*. It shows that although *mental models* are the deepest and the hardest to reach layer, the part that influences human decisions driving the *architecture* is learnable and adaptable. Using the cake analogy, this is saying that the part of what the original baker was thinking that influenced her new cake recipe decisions is possible to deduce, just like it is possible to deduce the architecture from the visible part of the problem. The Learning Principle states:

Learning Principle: (Common) *Mental models* influencing human *decisions* about *architectures* are *learnable* and *adaptable*.

How does the *Learning Principle* help?
This Systems Thinking Principle gives us the confidence that spending effort to dive deeper to understand the *mental models* that created the *architecture* shall, in principle, be possible and enable the highest leverage for our actions.

Why do we need the *Learning Principle* to guide our thinking and actions?
As with real diving, going deeper requires more capability and effort, but it does not imply that we shall keep diving forever! In real diving, we do not always dive to the lowest possible depth, every time that we dive. Yet the principle tells us that if we do, the deeper understanding has a higher leverage. The *SPs* are all about balancing effort and focus. So this principle teaches us to dive to the deepest level that is practical or justifiable for the complex challenges we face. It also teaches us that human interference with architecture comes from mental models. The word (*common*) at the beginning of the *Learning Principle* suggests that sometimes it is helpful to go even deeper: i.e., find the culture that influences the *mental models* that influence human *decisions* about the *architecture*.

5.10 The *Human Principle*

***Human Principle*: *Humans* impact and are impacted by the (wanted/unwanted) *emergence* of *all* reachable systems.**
This principle recognizes that *humans* have significant cause-and-effect relationships with *emergence*. On the one hand, they are a major contributor to wanted and unwanted emergent behavior and complexity of real-world problems. On the other hand, they are always a significant stakeholder and are impacted by *emergence*. So this principle recognizes that *humans* and *emergence* can create reinforcing and balancing loops that can be beneficial, harmful, or both. This remains true even if one or all of the *three systems* are partially or fully autonomous.

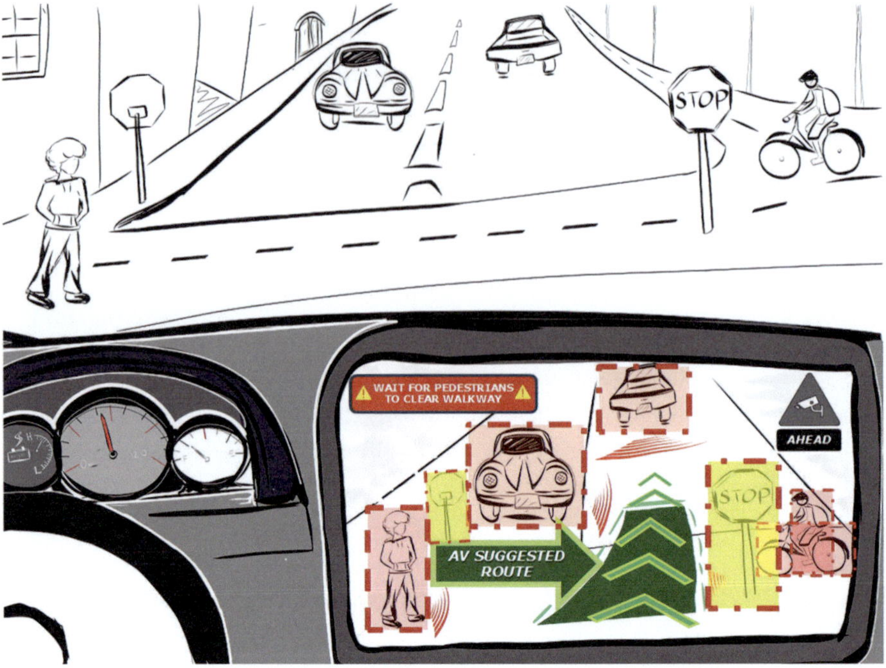

Fig. 5.15 *Human Principle*: *humans* impact and are impacted by the (wanted/unwanted) *emergence* of *all* reachable systems. Humans are always involved and significantly influence the interactions within and between the *three systems*. This remains true for autonomous systems, too

Simple examples of the *Human Principle*

> *Sometimes it seems as if there are more solutions than problems. On closer scrutiny, it turns out that many of today's problems are a result of yesterday's solutions.*— Thomas Sowell

Figure 5.15 shows a car with fully autonomous steering software. Part of its task is to identify all the obstacles in the driving path. Today's technology is capable of identifying cyclists and pedestrians with high reliability. But it has difficulty identifying their purpose or predicting what they will do with high reliability! Why? Similar to the iceberg analogy, mental models are deeper below the surface and it's harder to identify those than what is directly measurable from the camera image. Cyclists and other drivers with manual cars are not inside the SOI from the perspective of the car manufacturer, but they do interact with the SOI with the potential to cause severe unwanted

(continued)

behavior of the SOI. Moreover, users of fully autonomous vehicles express new behaviors as they learn to engage with the nuances of autonomous systems.

A true recent story of a proud new owner of an EV with a fully autonomous self-driving: After the cameras were calibrated on day 1 of purchase, the driver enabled the full self-driving mode in rush hour around 4:30 pm inside the city of Fort Myers. In the middle of a left turn at a busy junction, the light turned yellow, and the EV decided to immediately stop before entering the junction to wait until the turn signal was green again. Unfortunately, the driver of a manually steered truck behind had a different mental model: namely, that at least a few cars could still make the left turn before the yellow light turned red. The full stop by the autonomous EV caused a train of cars behind to come to a screeching halt, missing each other by a couple of millimeters. The drivers affected were extremely angry about the near miss and missing the light! The EV had properly identified the truck following behind. It also followed what might be deemed as safe behavior by not rushing through the yellow turn signal. Yet the added complexity was not knowing about or incorporating the mental models of the other drivers on the road. When confused, autonomous vehicles offer a manual override, but the whole scenario played out too rapidly for the EV driver to override. This story illustrates two human-caused unwanted behaviors leading to potentially catastrophic behavior: (1) The truck driver wanting to push through the yellow turn signal and (2) the EV owner excited about his new purchase who tried the new self-driving capability in a rush hour for the first time, before having enough experience with the EV to form a proper mental model of how it operates.

Figure 5.16 shows a "hype" curve that is a way to show the life cycle of new technologies through its "hype" stages. It shows a number of new and upcoming innovations on the horizon. Note many of these have a significant level of automation and some are even autonomous. Yet "all" these technologies have a potential for human-caused or human-involved beneficial and hurtful emergent behavior, regardless of their level of maturity and automation. Note that a number of technologies on the hype curve are safety-critical: i.e., they can cause damage or loss of human life so the *Human Principle* becomes a critical consideration for those innovations. Introducing new elements or operations without mapping the impact on and from human behavior can be potentially devastating.

Human Principle: *Humans* impact and are impacted by the (wanted/unwanted) *emergence* of *all* reachable systems.

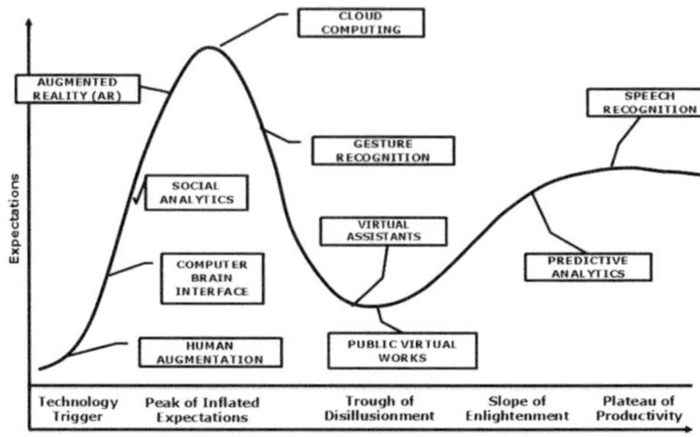

Fig. 5.16 *Human Principle*: *humans* impact and are impacted by the (wanted/unwanted) *emergence* of *all* reachable systems. All of the autonomous or semiautonomous technologies shown in various stages on the "hype" curve are not able to escape the impact of human factors or human interactions with the system, causing potentially unwanted emergent behavior. So, even autonomous systems cannot escape the *Human Principle*

How does the *Human Principle* help?
This principle urges us to consider all the relevant human influences in addition to the ways humans are impacted by our solutions. Looking at the economic, geopolitical, and environmental problems worldwide, we may notice or suspect a repeatable pattern that we the humans and our creations are creating problems at a faster rate than our ability to solve them. In systems thinking, the *Human Principle* is not intended to be a political statement or a complaint. It is intended to increase the awareness of systems thinkers as to how humans contribute to complexity in all reachable systems, regardless of whether any or all of the *three systems* are man-made or fully autonomous!

Why do we need the *Human Principle* to guide our thinking and actions?
The *Emergence Principle* described earlier helps the systems thinker become aware of (wanted or unwanted) complex behavior that emerges from the interaction or feedback between the three systems: *context system*, *system of interest* (SOI), and *enabling systems*. A way to anticipate this complexity is to identify the *3-systems* boundaries together with all the relevant interconnects and feedback between them. This principle, however, does not yet specifically address how to anticipate the complexity that is due to humans.

The *Human Principle* tells us that even if we have already identified the *3-systems* boundaries together with all the major interconnects and feedback

(continued)

between them, we still need to expect that some or all of the feedback can directly involve humans or be significantly influenced by them. In other words, we should always include human mental models, needs, considerations, factors, actions, impact, and influence when we try to understand *emergence*, regardless of whether one or more of the *three systems* are fully autonomous.

Sections 5.11 to 5.13 are included for completeness and not essential for beginners. Please move on to Section 5.14 upon first reading.

5.11 *Developing the Systems Principles*

The *SPs* of Sections 5.3 to 5.10 were extracted and distilled from a large body of systems thinking literature. The best known introductory material are listed in Section 5.13 for further reading. To develop the *SPs*, we used an iterative process with the following major elements:

- Statistical cluster analysis of survey results in 2008 from 700+ respondents distributed in 40 companies in Northern Colorado providing detailed insights into the nature of their complex systemic challenges (symptoms) and competencies (cures) that are likely to improve their odds of success in resolving them.
- Feedback from 300+ multidisciplinary students enrolled in a rigorous model-based systems thinking graduate class at CSU (SYSE 532) since 2009, ranging from just graduated to 25+ years of industry experience dealing with a large range of technical, semi-technical, and nontechnical complex challenges in government and multiple commercial and defense industries.
- Feedback from short week-long training on systems thinking to several aerospace and energy business units at Woodward, Inc., where each team (including both technical and nontechnical roles) applied the *SPs* to a domain specific but yet unresolved major complex business challenge.
- Feedback from 80+ students from various engineering, management, or business backgrounds enrolled in *CSU SYSE 505 Systems Thinking for the Real World* class. The main focus was on the application of *SPs* to real-world complex problems, aligned with the Bloom's taxonomy (https://en.wikipedia.org/wiki/Bloom%27s_taxonomy), where the students learned to:

  [CLO1] Classify failures or fiascoes that are due to the lack of systems thinking or culture.

  [CLO2] Classify successes that are due to good systems thinking culture.

[CLO3] Use the necessary hard-to-soft transdisciplinary systems thinking skills and rigorous tools, such that their odds of success improve significantly when dealing with complex real-world problems.
[CLO4] Reflect on the key *SPs* involved in wide range of industrial case studies from aircraft, energy, automotive, program management, cybernetics, and cyber-physical domains.
[CLO5] Create a systemic path for better decision-making in their capstone, design, or research projects using rigorous systems thinking tools.

- Discussion and brainstorming with CSU Systems Engineering faculty members who have solved real-world complex challenges in their deep and broad transdisciplinary careers in academia and industry.

5.12 Specific *Requirements* for the *Systems Principle*

Researchers [3, 19] are looking for laws like the laws of thermodynamics or Newton's conservation of mass, momentum, and energy that can be used to measure the complexity of systems and how it changes. However, this research is in the early stages, and there is much disagreement among systems science researchers as to the definition of what is a complex system or complex problem.[1]

Currently, the science of systems does not appear to have yielded laws or a theory that will directly derive the *SPs* we need. There are however many excellent and reputable systems thinking books and articles that attempt to teach good concepts that help our systems thinking skills and competencies. Section 5.13 lists a few generic and domain-specific references that contain excellent material relevant to systems thinking and principles. The classical approach to learning systems thinking requires learning, reflecting, and applying disjointed material from a large body of systems thinking references. This approach has enabled experienced professionals to become good systems thinkers later in their careers. However, it typically requires a major investment of resources and real-world application exposure to come up the learning curve. We do not believe that this is a practical path for young fresh talent or professionals trying to add systems thinking to their skill sets for the first time.

To provide a more practical path for our target audience of all disciplines, our method derived the following requirements for the *SPs* to comply with.

[1] For example, the Boeing 737 Max aircraft or its MCAS subsystem is considered by some as just "complicated" not "complex" because its operation can be reliably predicted like a watch. However, others call the same system or subsystem complex because its operation was not predictable in the presence of a combination of simple anomalies that combined to create very unpredictable adverse behaviors. In this case, a mis-calibrated angle of attack sensor worth circa $5000 not only caused two fatal crashes but almost brought down an empire costing more than $50 billion in damages with knock-on effects to other Boeing aircraft programs that do not even use the MCAS subsystem. This case was discussed in detail in Chapter 3.

The *SPs* shall with (minor exceptions):

Requirement-1. Be *universally true* for all systems.
Requirement-2. Be *universally applicable* to *complex/complicated* problems.
Requirement-3. Be *usable* by *technical and nontechnical* problem-solvers.
Requirement-4. Be useful in *isolation* or *together* with the other *SPs* in this book.
Requirement-5. Not *adversely impact* solving simple problems.
Requirement-6. Not use *names or concepts* that have a different standard meaning elsewhere in basic sciences, engineering, and management disciplines.
Requirement-7. Not use *names or concepts* that conflict with terminology in basic sciences, engineering, and management disciplines.
Requirement-8. Enable *classifying/understanding* systemic successes and failures.
Requirement-9. Form the *basis for rigorous tools* that apply systems thinking.
Requirement-10. Apply equally well to *system of systems*.
Requirement-11. Be *understandable* and *usable* by the target *audience*.[2]
Requirement-12. Have a meaningful easy-to-remember *label*.

Requirement 1 ensures that the truth contained in the principle applies to all types of systems. Whether it is a system that implements a homelessness policy or an electromechanical actuation system, the truth of the principle shall guide the correct thinking, behavior, or action. *Requirement 2* ensures that the principle guides the correct thinking, behavior, or action when facing complex and complicated problems. *Requirement 3* ensures that both technical and nontechnical users are catered for equally. *Requirement 4* ensures that there is no conflict or confusion among the *SPs* so that they can be used alone or together depending on the needs of the problem in hand. *Requirement 5* ensures that using the *SPs* will not lead to bad solutions when the problem is simple enough so that linear thinking would have worked. *Requirements 6 and 7* avoid inadvertent domain-specific guidance by not using names or concepts that are similar or conflicting with domain-specific ones. *Requirement 8* ensures that the *SPs* can be used to evaluate, classify, and understand the deeper less obvious causes of success and failure.*Requirement 9* ensures that the *SPs* are strong enough to form the basis for the rigorous tools or that they can help evaluate whether a rigorous tool is implementing systems thinking. *Requirements 10, 11, and 12* ensure that the *SPs* are generically applicable.

[2] Targeted audience here is secondary, undergraduate, and graduate students of all technical and nontechnical disciplines.

5.13 Relevant *Systems Literature*

There is a vast amount of literature on systems thinking and related principles. This section presents our interpretation of the "essentials" for beginners to intermediate systems thinkers if the abstraction, level, style, language, details, examples, and organization of the material in this book are not a good fit:

Thinking in Systems: A Primer (Meadows) [12] is the ultimate beginner's guide to understanding how the world works through the eyes of the systems thinker. It breaks down the structure and behavior of complex systems into bite-sized pieces that anyone can grasp. It's a must-read for anyone wanting to dive into systems thinking because it lays the groundwork for all the systems literature that followed.

Systems Thinking Basics: From Concepts to Causal Loops (Anderson et al.) [1] This is also great for beginners of all backgrounds, starting with simple principles and moving on to tools like causal loop diagrams, etc. It is a nice transition or bridge to the heavier material in Sterman's book next.

Business Dynamics: Systems Thinking and Modeling for a Complex World (Sterman) [20] This is the advanced systems thinking course we typically take after mastering Meadows' (and Anderson's) book. It gets into the nitty-gritty of how to model systems using system dynamics, especially in a business context. It is more technical than Meadows and builds on her foundation by giving us the hands-on tools to apply systems thinking to real-world problems. Meadows gives us the basic concepts, and Sterman shows us how to put those concepts into action. They're both essential reads for anyone serious about understanding how systems work and using that knowledge to make better decisions.

The Fifth Discipline: The Art and Practice of the Learning Organization (Senge) [17] This is all about building "learning organizations" using the principles of systems thinking, where people are constantly growing and working together to achieve shared goals. He dives into how things like teamwork, personal growth, and shared visions can make companies way more successful and adaptable, and it's considered a classic in organizational learning.

A Primer for Model-Based Systems Engineering (Long et al.) [9] This provides a comprehensive introduction to the engineering practice of model-based systems engineering (MBSE). It emphasizes the use of formal models to support the requirements, design, analysis, verification, and validation of engineered complex systems throughout their lifecycle. While *Thinking in Systems: A Primer* offers a foundational understanding of Systems Thinking Principles and *Business Dynamics: Systems Thinking and Modeling for a Complex World* presents the application of system dynamics modeling, this book focuses specifically on the practical application of MBSE in engineering disciplines. It provides a structured approach to using models as a central artifact in the systems engineering process, promoting a more rigorous and efficient approach to system development.

Industrial Dynamics (Forrester) [4] published in 1961, pioneered the application of feedback control theory to understand the complex behavior of industrial systems. It introduced key concepts like stocks, flows, and feedback loops and emphasized the use of computer simulation for analyzing these systems. This groundbreaking work laid the foundation for the field of system dynamics, influencing not only industrial management but also many other fields like economics (Sterman) [20], social sciences (Stroh) [21], and environmental studies (Meadows) [12].

World Dynamics (Forrester) [5] and *Limits to Growth: The 30-Year Update (Meadows et al.)* [13] Building on *Industrial Dynamics*, his 1971 book *World Dynamics* expanded system dynamics modeling to global systems, examining the interplay of population, resources, pollution, and economic growth. This work directly inspired *Limits to Growth: The 30-Year Update* (1972), which used a similar model to explore the potential consequences of exponential growth on a finite planet. Both books, published during a period of growing environmental awareness, sparked widespread debate about sustainability and the long-term trajectory of human civilization. They remain influential in discussions about global challenges and the need for systemic change.

Systems Thinkers (Ramage et al.) [15] offers a unique biographical approach to understanding systems science, providing intimate portraits of 30 key figures (including a number of authors in this list) who shaped the field. Unlike most books on the disciplines, which focus on theories, this book delves into the lives and motivations of the thinkers behind the theories and practices, highlighting the human element behind the concepts and revealing how their personal experiences influenced their work.

Mental Models: Towards a Cognitive Science of Language, Inference, and Consciousness [8] (1983) significantly advanced the concept of mental models. This work laid the foundation for the use of mental models in contemporary systems thinking, particularly within the "iceberg model." The iceberg model [17, (1990)] uses the metaphor of an iceberg to illustrate different levels of understanding within a system. Mental models, residing at the deepest level of the iceberg, are seen as the underlying beliefs and assumptions that drive the observable patterns and events. By surfacing and challenging these mental models, systems thinking aims to foster more effective problem-solving and systemic change.

Principles of Systems Science (Mobus et al.) [14] provides a comprehensive and rigorous treatment of systems science, emphasizing formal analysis and practical application. Systems science itself provides a broad theoretical framework for understanding complex systems, encompassing (among many other disciplines) both ST and SD. ST itself focuses on qualitative aspects like interrelationships and mental models, and SD adds a quantitative layer by using tools like simulation modeling to analyze system behavior over time.

Feedback Thought in Social Science and Systems Theory (Richardson [16]) is a deep dive into the history and philosophy behind the core concept of systems thinking: feedback loops. He explores how this idea of cause-and-effect circling back on itself has appeared in different fields over time and how it shapes our understanding of social systems today. It's a more academic book than Meadows or Senge.

Critical Systems Thinking: A Practitioner's Guide (Jackson [7]) Jackson has two other easier-to-follow books on systems thinking and is regarded as a reputable source. This recent book maps systems thinking regarding history, science, key names, theory, and disciplinary practice of systems thinking. It also acts like a good grammar book that covers a wide range of vocabulary related to ST. Children do not learn to speak or walk by learning heavy grammar or differential equations that capture their motion first. They learn by doing first and then proceed to fine-tune and improve their doing with theory, science, laws, rules, etc. By analogy, fresh young talent or disciplinary professionals coming into ST for the first time will not know how to think or approach problems differently just by reading this book. However, this is an excellent reference for those who can already walk the talk of ST and now want to know where ST comes from and how it is related to everything else.

We do not claim that the above list of references, or even this book, is the best. Our main motivation to include this list is to create a path to learn more or to simply get another perspective on *Systems Principles*.

5.14 Reflection on Chapter 5 Material

Part I of this book gave us introductory training so we would have the basic knowledge and terminology to understand and absorb the teachings of this chapter. In this chapter, we introduced our *Generic Systems Thinking Principles* or *Systems Principles* or *SP*s. This chapter kicks off the philosophy part of this book because the principles are the core. The *SP*s are the keys to understanding and influencing real-world problems. We explained each *SP* with simple real-world examples to highlight its meaning and subtleties. We also showed the usefulness of each *SP* and why we need it to guide our actions and behaviors.

Next, we learn to develop an eye for the *SPs* in the real world around us (Chapter 6), followed by how we can systematically apply the *SPs* to improve our odds of success when we attempt to influence the system (Chapter 7).

References

1. Anderson, V., & Johnson, L. (1997). *Systems thinking basics: From concepts to causal loops*. Pegasus Communications. ISBN: 978-1883823122.
2. Borky, J., & Bradley, T. (2019). *Effective model-based systems engineering*. Springer.
3. de Weck, O. (2023). Keynote: Conservation of complexity: The first law of system science. In *International Conference on Engineering Design (ICED) in Bordeaux, France*.
4. Forrester, J. W. (1961). *Industrial dynamics* (464 p.). MIT Press.

5. Forrester, J. W. (1973). *World dynamics* (2nd. ed., xv, 144 p.). Wright-Allen Press. ISBN: 0262560186.
6. International Council on Systems Engineering. (2023). *Systems engineering handbook: A guide for system life cycle processes and activities: Fifth edition*. Wiley.
7. Jackson, M. C. (2024). *Critical systems thinking: A practitioner's guide*. Wiley. ISBN: 9781394203574.
8. Johnson-Laird, P. N. (1983). *Mental models: Towards a cognitive science of language, inference and consciousness* (xiii, 513 p.). Cambridge University Press. ISBN: 0521241235 0521273919 (pbk.).
9. Long, D., & Scott, Z. (2011). *A primer for model-based systems engineering*. Lulu. com. ISBN: 1105588106.
10. Madni, A. M. (2018). *Transdisciplinary systems engineering: Exploiting convergence in a hyper-connected world* (1st ed.). Springer.
11. Madni, A. M., Augustine, N., et al. (2023). *Handbook of model-based systems engineering* (1st ed.). Springer.
12. Meadows, D. H. (2008). *Thinking in systems: A primer*. Chelsea Green.
13. Meadows, D. H., Randers, J., & Meadows, D. L. (2004). *The limits to growth: The 30-year update* (xxii, 338 p.). Chelsea Green Publishing Company. ISBN: 1931498512 (hardcover alk. paper) 193149858X (pbk. alk. paper).
14. Mobus, G. E., & Kalton, M. C. (2015). *Principles of systems science*. Understanding Complex Systems (xxvi, 755 p.). Springer. ISBN: 9781493919192 (hardback) 1493919199 (hardback) 9781493919208 (ebook) 1493919202 (ebook) 1860-0832. https://doi.org/10.1007/978-1-4939-1920-8.
15. Ramage, M., & Shipp, K. (2009). *Systems thinkers* (ix, 316 p.). Springer. ISBN: 1848825242 (pbk. acid-free paper) 9781848825246 (pbk. acid-free paper).
16. Richardson, G. P. (1991). *Feedback thought in social science and systems theory*. University of Pennsylvania. ISBN: 0812213327.
17. Senge, P. (2006). *The fifth discipline: The art and practice of the learning organization*. Random House Books.
18. Shahroudi, K. E., et al. (2024). Empowering real world complex problem solving: Socio-technical applications of model-based systems thinking (MBST). In *34th Annual INCOSE International Symposium (IS2024), Panel Presentation*.
19. Sinha, K., & de Weck, O. L. (2016). Empirical validation of structural complexity metric and complexity management for engineering systems. *INCOSE Systems Engineering, 19*(3).
20. Sterman, J. D. (2000). *Business dynamics: Systems thinking and modeling for a complex world*. Irwin McGraw-Hill.
21. Stroh, D. P. (2015). Systems thinking for social change: Making an explicit choice. *Reflections, 14*(3), 35–42.

Chapter 6
Seeing Systems Principles Around Us

Abstract Chapter 5 introduced us to the systems principles. However, before we start learning their application in Chapter 7, it is helpful to start developing eyes for systems thinking in the world around us, such as in music, cartoons, games, sports, and movies. Evidence of system principles and cultures of systems thinking are already in play around us. Seeing how systems thinking plays out in everyday examples will make the *systems principles* much more tangible. Metaphors are a well-known method for effectively communicating complicated information. Let's take a moment to look around! Can we spot examples of different systems principles at work by considering what engages us regularly? How did the country pop genre evolve? Why are South Park episodes so clever and funny? Why do we experience wildly different results when we play multiplayer games online? Where does the versatility of a deck of cards come from? What makes the heroes in action and adventure films memorable besides the stunt doubles and big budgets? What is the journey that pushes some to move to the dark side? Discovering the systems principles at play in the world around us is fascinating and a great way to learn how things connect and influence each other!

Chapter 5 introduced us to the systems principles. However, before we start learning their application in Chapter 7, it is helpful to start developing eyes for systems thinking in the world around us, such as in music, cartoons, games, sports, and movies. Evidence of system principles and cultures of systems thinking are already in play around us. Seeing how systems thinking plays out in everyday examples will make the *systems principles* much more tangible. Metaphors are a well-known method for effectively communicating complicated information. Let's take a moment to look around! Can we spot examples of different systems principles at work by considering what engages us regularly? How did the country pop genre evolve? Why are South Park episodes so clever and funny? Why do we experience wildly different results when we play multiplayer games online? Where does the versatility of a deck of cards come from? What makes the heroes in action and adventure films memorable besides the stunt doubles and big budgets? What is

K. Eftekhari Shahroudi et al., *Practical Systems Thinking*,
https://doi.org/10.1007/978-3-031-89394-0_6

the journey that pushes some to move to the dark side? Discovering the systems principles at play in the world around us is fascinating and a great way to learn how things connect and influence each other!

6.1 Country Pop: An Evolved Genre System

A genre is a typical, recognizable, repeated combination of form, content, function, and situation. When genres intersect, the term given to the intersecting genres that facilitate a particular kind of work is "genre system" [8]. This genre system evolution highlights the principle *evolution principle*: systems have *life cycles* and they *evolve*. The particular genre system we will analyze here is country pop.

Defining the characteristics of the pop music genre is challenging because it constantly evolves. "Pop" is short for "popular," which can have a comprehensive definition. If we look at music that was popular a century ago, it would be vastly different from what is popular and influential today in the twenty-first century. Two standard features of the pop music genre that seem to be holding true over time are as follows:

- Has a hook: A song hook is a short lyrical line or melodic phrase used to catch the listener's ear and make a song engaging [5]. Some famous pop song hooks include "I'm a mess, I'm a loser, I'm a hater, I'm a user" (Bebe Rexha, 2018), "I will survive" (Gloria Gaynor, 1978), "You ain't nothin' but a hound dog, cryin' all the time" (Elvis Presley, 1956), and "If you liked it, then you should've put a ring on it" (Beyonce, 2008).
- Contains a tempo that we can easily dance to: Not all music is easy for us to dance to. A song that gets us instantly and spontaneously moving without overthinking is a song that can easily be danced to.

If we listen to the 1970s, 1980s, and 1990s pop music, we will notice that they are characteristically different and that the pop music genre has evolved over the decades. We can all agree also that a song has elements (hooks, tempo, melodies, instruments, lyrics, and voice) interconnected to create a feeling, emotion, or awareness of a topic the artist is concerned about. So, even a song itself fits the description of a system given in Chapter 1.

On the other hand, the country music genre is slightly easier to define. Country music is defined as "a style and genre of largely string-accompanied American popular music having roots in the folk music of the Southeast and cowboy music of the West, usually vocalized, generally simple in form and harmony, and typified by romantic or melancholy ballads accompanied by acoustic or electric guitar, banjo, violin, and harmonica" [1]. Country music is also widely known for its focus on storytelling.

The typical stages of the life cycle of a song [4] regardless of genre are as follows:

- Incubation: determine a vision for the song—what feel do we want the song to elicit, what message are we trying to get across, what audience are we trying to reach, and why—and come up with a plan for executing the rest of the stages
- Writing: determine verses, chorus, and hooks
- Recording: capture tracks
- Editing: edit each track to contain the best possible performance
- Mixing: blend individual tracks to make a solid, cohesive song [2]
- Mastering: quality control for the song to catch and resolve any issues, put the final polish on the song
- Release: distribute and promote the song
- Profits: generate revenues from albums, streaming, movies, and media
- Retirement: lower popularity and revenue generation since the initial launch

The genre system of country pop is characterized largely as string-accompanied songs with straightforward stories that have a hook and an easily danceable tempo. Since a genre system is characterized by links between context, text, and activity, let us break down a commonly known country pop song, "Man! I Feel Like a Woman" by Shania Twain [3], and show how changes to the typical life cycle of this song system influenced the evolution of the country pop genre system.

- Context: In writing the song, Shania was inspired by her childhood experiences. The storytelling element of the song and her history as a country music artist makes it fall into the country music category.
- Text: The song arguably has more than one hook. It starts with an enthusiastic "Let's go, girls," and the chorus's hook is the song's title, "Man! I feel like a woman." These hooks are one of the reasons this song evolved from a country song to a country pop song.
- Activity: Another reason for the evolution was the intentional editing and mixing Shania and her music producer did to give it a danceable tempo [6].

Eventually, enough songs were made with the characteristics of both pop and country music genres justifying the need for the new genre system of country pop.

This is an example of how systems can evolve when changes, intentional or not, are made during the various stages in an associated life cycle.

Circling back to a song as a system, is it possible for there to be significant changes or new events for an individual song after "retirement"? The answer is Yes! Like the biological species on earth, songs and genres continue to evolve, morph, and regenerate. The Soul Album by Seal, released in 2012, was a hit in ten countries where he re-sang many songs that were hits decades earlier, such as Stand By Me by Ben E. King. Dua Lipa, a current pop singer in hot demand, has at least ten songs that have been streamed more than a couple million times each. She told *Variety* that she was inspired by *Pink* when she was little. However, their genres are not perfect matches. Their songs both technically fall into the pop genre category, but Pink is more rock-pop and Dua Lipa is more electro-pop. Songs and music genres do not just disappear; they evolve and give birth to new generations through various artists or by influencing the incubation stage of future songs.

Now, would we have been able to predict that songs evolve beyond the retirement stage? The answer is yes because a song is a system, and the *evolution principle* explained in Chapter 5 states:

Evolution principle: systems have *life cycles* and they *evolve*.

The *evolution principle* also explains why defining the characteristics of the pop music genre is difficult. Chapter 5 explained that all the systems principles apply all the time but that sometimes one provides a more direct or easier path to understand a real-world issue. For example, when the rock and roll genre first emerged in the 1950s, it was at odds with the mental model of large sections of society who thought it was rude, obscene, vulgar, disgusting, and anarchist. Many did not like the sound and the rhythm, while others did not like the energetic dance of partners or the socializing of segregated races in the same environment. The dominant mental models were against the whole *architecture* (elements, interconnect, and purpose) of rock and roll. However, it soon became an activity that young people of many races and nations enjoyed together. Over time, the mental models of rock and roll adapted, and it became a dominant best-selling genre. How would we have predicted that the mental models of a large section of society about rock and roll might change? The answer comes from the *learning principle*:

Learning principle: (common) *mental models* influencing human *decisions* about *architectures* are *learnable* and *adaptable*.

6.2 South Park Crudely but Perfectly Captures the Human Principle

In the South Park episode "Giving Real Estate a Try" shown in Fig. 6.1, they play on the fundamental problems the real estate boom in Denver is causing for front-range and mountain communities. Sure, South Park can be vulgar, but don't "throw the baby out with the bathwater" here; this episode is an excellent example of the *human principle*: *humans* impact and are impacted by the (wanted/unwanted) *emergence* of *all* reachable systems. In real life, Colorado faces a shortfall of 100,000 homes and

Fig. 6.1 South park episode "Giving Real Estate a Try" https://southpark.cc.com/video-clips/u7qkru/south-park-giving-real-estate-a-try

apartments, the second worst deficit of any state after California, according to a study from Up for Growth. A home buyer misery index from the Common Sense Institute found households in the state's largest counties facing record-high levels of stress [11].

So how did it all start? Thousands of people wanted to live in Colorado for all sorts of reasons: the lure of "more" affordable housing, especially if they come from a less affordable state or area, good job options, decent weather year-round, easy access to snow sports, easy to travel in and out, etc. In summary, Colorado is pretty awesome, so it makes sense that people want to live there. Individuals who sell houses are known as real estate agents. The episode starts with real estate agents in South Park enthusiastic about the trend of city people wanting to move to the country. These real estate agents conclude that the problem is that they do not have enough real estate agents to keep up with demand. This is a more convenient conclusion than determining that a shortage of houses is the issue because they can hire real estate agents more quickly than build houses. So, they make the wise decision to hire Cartman's mother. Two subtle but interesting things happen at this point—they highlight the low barrier to entry for someone to become a real estate agent, and Cartman's mom says she needs the job because her rent is increasing with all the new people moving to town.

To be a successful real estate agent in South Park, they must get "their picture taken, get people to like their picture, and get people to like them when they show people the house." In real life and most states, getting a license to help people buy or sell a home requires only a few hundred dollars, several weeks of coursework, and a passing grade on a multiple-choice test [10]. We all would love to make lots of money for as little work as possible! This surplus of real estate agents puts "buyers and sellers at risk of ending up with dubious advice on one of the biggest transactions of their lives, while the capable agents are forced to spend inordinate amounts of time and money trying to stand out from the pack. They also waste hours dealing with incompetent colleagues as they try to get a sale to the finish line. When too many agents are fighting over too few deals, they're motivated to keep commission rates high so they can continue to eke out a living—to the detriment of consumers" [10]. One of the ways real estate agents in South Park try to stand out from the crowd is through fancy videos of the properties. The cost to make these videos may initially be paid by the real estate agent, but, make no mistake about it, it will be made up for in their commission.

The problem with increasing rent is due to the decrease in their supply—houses, not real estate agents. Landlords increasing rents cause another problem. In Colorado specifically, "there is a shortage of rental homes affordable and available to extremely low-income households (ELI), whose incomes are at or below the poverty guideline or 30 percent of their area median income (AMI). Many of these households are severely cost burdened, spending more than half their income on housing" [9].

Initially, the local government and business owners of South Park are enthusiastic about all the economic benefits that "city people" bring. Then, there are too many city people. Eventually, they realize that if too many city people move to South Park, their town will become a city, and since "the one thing city people hate most is other city people," the boom will be short-lived when the city people move on. They conclude by blaming the real estate agents for the abundance of city people.

6.3 A Basketball Game: Where Emergent Behavior Abounds or Rebounds

A basketball game as depicted in Fig. 6.2, is where emergent behavior abounds or rebounds! Watch a basketball game at any level to see an excellent example of *emergence principle*: *interactions* (including feedback) within and between the

Fig. 6.2 A basketball game highlights the *emergence principle* where the synergy between the players is more important than the quality of individual players

3-systems cause (sometimes unwanted) *emergent behaviors*. This principle is easily seen in action if we watch any sport and know what to look for. Let's break down basketball. Basketball is a game where two teams of five players each take turns maneuvering the basketball down a court into one of the two hoops at opposing ends. At the start of the game, each team is assigned a side, and after a defined period, they change sides. The goal for each team is to make as many baskets as possible to win. Basketball is all about working as a team to create opportunities for passing, scoring, and supporting each other to prevent the opposing team from doing the same. First, let's agree that the system of interest is the specific game played between the two teams; the context system includes the physical court and the league rules, and the enabling systems include the individual team systems. Multiple feedback loops between these systems during the basketball game cause emergent behavior.

Changes to the structure of the enabling systems of the team can cause emergent behavior or changes in the overall strategy. The other teams' winning strategy may change depending on the perceived skill of the individual currently playing point guard. As the game progresses, new strategies will emerge as each team learns the other's strengths, weaknesses, and early strategies. This highlights the *learning principle*: (common) *mental models* influencing human *decisions* about *architectures* are *learnable* and *adaptable*. For example, in an attempt to make a shot, Jack passes the ball to Jason, who attempts a three-point shot. Depending on whether he makes the shot or not, feedback is generated. If the attempt is unsuccessful, the teammates can try to get the ball back, and if they are successful, they go back to guard their side. Additionally, based on how the team performed, the coach can advise the team to make necessary adjustments to increase their chances of success. Actions lead to subsequent actions; these feedback loops are constant and dynamic in a basketball game.

Players' decisions during the game are based on their skill and ability to react to the changing position of the basketball around the court. Each player must balance effort and focus by dynamically reading multiple players' perspectives to ascertain the best course of action. They must predict which of their list of potential actions has the least chance of risk with the greatest chance of gain and when more significant risks for greater potential rewards must be taken. They must watch for win/lose mindsets; setting up a pick for a teammate may be a great help, but if the utilization of the pick leaves the teammate covered by multiple opposing players, the teammate's options become more limited. Most notably, players must see themselves as part of the team system because a one person show rarely carries the team to victory. The victor in a basketball game is often the team with the best teamwork capabilities. While it can be helpful to have a superstar player to win, it is still a five-person team, and the team is worth more than just the sum of its parts. A player's value can increase through interacting with others (passing, setting picks, positioning, anticipating), and this is not just based on individual skills. In addition, a basketball team with star players can quickly lose ground to a less talented team if they are not playing with purpose and structure. Not only is practice essential for improvement through the development of structured offenses and defenses, but it is also essential to play with purpose. This represents the desire to win and to achieve the best performance possible by structuring behaviors and, thus, outcomes.

A shift in momentum due to these feedback loops is a classic example of emergent behavior in sports. A fantastic play in basketball can reinvigorate one team and discourage the other. Team flow is another example of emergent behavior that happens as a result of multiple positive feedback loops. Team flow is when team members trust one another to do their jobs, and plays are executed more instinctively than calculatedly.

6.4 Online Gaming: Masters of Systems Thinking

Online multiplayer games with an example shown in Fig. 6.3, exemplify the *holism principle*: (complex) *systems* are (much) *more* than the *sum* of their *parts*, *elements*, *interconnects*, and *purpose*. And *boundaries principle*: *systems boundaries* depend on *perspectives*, *purposes* and *roles*. Let's examine how these games highlight these systems thinking principles.

In games like the World of Warcraft, players earn experience points and gain additional capabilities as their character levels increase. Players can take on more complex challenges with each level and gain more experience points. The rate at which players gain these points increases with each level, so there is a reinforcing feedback loop where success leads to more success. The reinforcing loops encourage more gameplay. Game creators also introduce balancing feedback loops to keep players from being overpowered. For example, in Fortnite, higher-level characters are infrequently placed with lower-ranked players to keep the game fair and challenging. In games without these balancing loops, higher-level characters have higher kill rates and go on to dominate gameplay. In online multiplayer games,

Fig. 6.3 An online multiplayer racing environment utilizes synergy among the players so that unique outcomes and experience *emerge*

the mechanics, players, environment, and rules are all interconnected, leading to complex behavior.

In the Battlefield series of games, the teaming, direction, and strategies are all driven by players. Individuals or groups can affect alliances, priorities, and directions across the game world, directing each round of gameplay. In the game Don't Starve, players must decide how to gather resources, interact with roving creatures, and manage their physical and mental health. During winter, food sources become scarce, and players must plan by stockpiling resources. Choosing to utilize resources too quickly can cause starvation, which can cause players to hunt more aggressively or eat spoiled food, affecting a player's health and showing how one action affects multiple aspects of the player's experience.

Sid Meier's Civilization series involves managing resources or stocks and understanding how these stocks change over time. Losing track of these resources can affect gameplay. Players must balance the growth of their civilization by effectively managing the stocks and flows in their world. Players change the rate of the flows into their stocks by investing in technology, education, and other system mechanics representing subsystems. Players can also modify the structure of gameplay by developing and expanding societal belief systems. Changes in this structure consequently affect how resources are produced or used. Unique system behavior is possible with each instance of an online multiplayer game. The more players and interactions an online game presents, the more players' collective action creates emerging behavior.

In Minecraft, players use fundamental building blocks to create detailed environments that affect how players interact. In the Sim series of games, Sim characters interact with each other based on their personalities, needs, and the player's actions, creating unique and unexpected social outcomes. In multiplayer online battle games, teams of players are matched to compete against other semi-randomly matched teams. Each match is unique due to the combination of the player's skills, chosen strategies, experiences, and decisions. Synergies emerge between players, creating an emerging experience where no two matches are alike.

These online game systems are much more than the sum of their parts, and the game's outcome depends on the collective players' perspectives, purposes, and roles.

6.5 Solitaire and Blackjack

Solitaire and blackjack shown in Fig 6.4 are two card games that exemplify the *architecture principle*: the *purpose*, *structure*, and *behavior* of systems *influence* each other. They both use the same deck of cards, but the purpose, structure, and behaviors of one another in the way that the deck of cards is used result in two wildly different card games. Let's do a quick review of both games.

Fig. 6.4 Solitaire and blackjack highlight the *architecture principle*: the *purpose*, *structure*, and *behavior* of systems *influence* each other. Same deck of cards, but very different experiences and outcomes

6.5.1 Solitaire

Purpose

Solitaire organizes a shuffled deck of cards into four foundation piles representing each suit in ascending numerical order as quickly as possible while adhering to game rules. Solitaire is a card game played alone, which makes it easy to play when we want. It is also a good game for keeping those cognitive skills sharp!

Structure

We start with a shuffled deck of cards. We create seven columns, with the first column on the left having one card, the second having two cards, and the third having three cards, continuing to seven columns. The top card in each column is turned face up, and the rest are face down. The remaining cards go face down in a single stack called the stock pile. There are four spots for the foundation piles. Cards that are face up and showing may be moved from the stockpile or the columns to the foundation stacks or other columns. Moving a card to a column must be one less in rank and the opposite color. For example, if it was a nine of hearts (red), we could put eight spades or clubs onto it. Stacks of cards may be moved from one column to another as long as they maintain the same order (highest to lowest, alternating colors). We can start a new column with a King if we get an empty column. Any new column must be started with a King (or a stack of cards that starts with a King). To get new cards from the stockpile, we turn three cards at a time face up into the stack next to the stockpile called the waist stack. We can only play the top card off the waist stack. If we run out of stock cards, turn the waist stack over to make a new stockpile and start again, pulling the top three cards off, turning them over, and starting a new waist stack.

Patterns of Behavior

We first move the cards on the table or pull them from the stockpile to see our options. We move cards to the foundation piles as soon as possible or wait to ensure we don't need them to move cards around the columns. We then decide whether to move a card from a column with the most face-down cards or move the King to an empty column.

6.5.2 Blackjack

Purpose

In blackjack, everyone plays against the dealer. Players receive cards face up or face down, depending on the context, while the dealer's first card is face up, and the second is face down. The game's object is to get closer to 21 than the dealer without going over 21. If a hand goes over 21, it is called a "bust," and the wager is lost. However, if the dealer busts, all players who did not bust during their turn win by default.

Structure

Unlike solitaire, blackjack requires at least two people–a dealer and a player. A specific blackjack table is used in a casino. If playing at home, the table is not necessary, but the orientation of the dealer facing the player should be the same. The rules of basic blackjack [7] are as follows:

- Players place their bets in their betting positions.
- The dealer gives each player two cards face-up in front of their respective betting positions.
- The dealer gives themselves two cards, one face-up and another face-down.
- Starting to the dealer's left, each player can draw more cards. They can either "hit" or"stand". They are given an extra card if they "hit". Players can "hit" as many times as they wish. If they "stand", they indicate they do not want the dealer to draw more cards for them. If the player busts at this point, they immediately lose their bet.
- The dealer must take cards until their total is at least 17. Once they reach 17 or above, their cards are totaled, and each player's hand is compared to the dealer's. However, if the dealer busts, players who did not bust win immediately. If the dealer does not bust, then each player who scores higher than the dealer wins. If they tie, then just the player's original bet is returned. If they score lower than the dealer, they lose their bet.

Patterns of Behavior

The art of deciding when to take a card as a player can vary significantly in complexity. One of the most significant variables is what card is shown as the dealer's face-up card. If the dealer's card is a seven or greater, then a player must assume the dealer will at least get a 17, and most players will hit until they reach at least 17. However, if the dealer has a 4, 5, or 6, these cards are known as "bust" cards, where the odds of the dealer busting go up. Many players in this scenario will take fewer cards to reduce their risk of busting since the dealer is more likely to bust. It is usually reasonably obvious for players when they need to hit, i.e., the total of their cards is less than 11. Still, players' behavior patterns change significantly in the gray area when the dealer has a bust card or the player's cards total anywhere between 15 and 17. The decision is based on individual appetites for risk and sometimes gut feelings. Experts play differently than novices. Experts have developed patterns for systematically determining if the odds are in their favor, whereas novices play based on gut feelings. Blackjack is a game of chance, but managing the complexity of interactions and understanding the more extensive system can significantly increase a player's chance at success.

6.6 Why the RDA in Avatar Became Evil

Action and adventure films give us access to larger-than-life experiences. They are built around our desire to search for "more" in our lives. The hero or protagonist in these films is a systems thinker who recognizes that to solve the challenging problem, they have to understand the broader dynamic system, aka the *3-systems principle*: *systems* come in (at least) *3s*, *context systems*, *systems of interest (SoI)*, and *enabling systems*.

On the flip side, the antagonist is shown as having a complete disregard for understanding the three systems. In many of these movies, the antagonist may begin implementing their system of interest for noble reasons. Still, their failure to take the time to fully understand the context and enabling systems and the resulting chaos that ensues drives them crazy. In these movies, that crazy feeling always turns into evil villainy.

Let's break down Avatar shown as depicted in Fig. 6.5, the science fiction film by James Cameron that was released in 2009. In Avatar, let's first consider the proposed system of interest, the system that the humans from Earth are setting up to mine the valuable mineral unobtanium on Pandora. Unobtanium is a fictional term used to describe a material ideal for a particular application that is difficult or impossible to obtain. Unobtain-ium, get it?! The system that the humans are setting up is called the Resources Development Administration (RDA). In 2154, humans completely depleted the Earth's natural resources, so apparently, people

Fig. 6.5 The Avatar story on planet Na'Vi is full of unintended and unwanted complex behavior created by the interaction of humans with a complex alien ecosystem. The unintended consequences can be explained by the *emergence* and *human* principles

were having trouble keeping the lights on and driving their flying cars. The RDA was arguably once a noble system developed within the context of these broken earth systems. Through much research, the RDA found that Pandora was rich in unobtanium, and the solution became clear—set up a mining operation on Pandora. Pandora is a context system since that is where the mining operation was to be set up. Two sub-context systems on Pandora provide challenges for the RDA. First, Pandora is already inhabited by the Na'Vi, a species of 10-foot tall (3.0 m), blue-skinned, sapient humanoids that live in harmony with nature and worship a mother goddess named Eywa. So, the RDA is trying to set up a mining operation among people who live in harmony with nature, which will be a challenge. Second, the atmospheric system on Pandora is toxic to humans, so an enabling system of avatars had to be developed for humans to explore the planet. The avatars are Na'Vi-human hybrids (grown from human + native DNA). Another enabling system is the employees who operate the avatars. The employees have to be a genetic match, so when one of the humans an avatar was made for dies, the RDA quickly recruits the identical twin Jake Sully (the protagonist in this story) to take his place operating the avatar.

So, spoiler alert: the RDA is unsuccessful. Without getting into the rest of the movie, just by identifying the three systems, the systems thinker can already see all the potential risks and issues the RDA would face to succeed. In the movie, they were not just unsuccessful; their lack of systems thinking also ultimately positioned them as evil, an unintended consequence no one wants!

Discussion question: what is an alternative ending to Avatar that would protect the Na'Vi people and Pandora while still helping humans on Earth? The proposed alternative ending shall explain how the context and enabling systems would be leveraged.

6.7 Reflection on Chapter 6 Material

Evidence of system principles and cultures of systems thinking are already in play around us. The previous Chapter 5 rolled out the systems principles (*SP*s) we need to understand real-world problems and find their best leverage points. This chapter helped us develop an eye for the *SP*s in the world around us using everyday sports, card games, computer games, and movie examples. Next, we move on to Chapter 7 to learn how to apply the *SP*s in our actions and behavior.

References

1. Austrew, A. (2019). *What Kind Of Songs Get To Be Called "Country Music"?* https://www.dictionary.com/e/what-does-country-music-mean/.
2. Avid. (2024). *Music Production: Guide to Producing & Releasing Songs*. https://www.avid.com/resource-center/how-to-produce-music.
3. Eames, T. (2024). *The Story of... 'Man! I Feel Like a Woman' by Shania Twain*. https://www.smoothradio.com/features/the-story-of/man-i-feel-like-a-woman-shania-twain-lyrics-meaning/.
4. George's Music. *The Recording Process Explained*. https://www.georgesmusic.com/Article-117.html.
5. ICON Collective. *What Is a Hook in a Song and How to Write One*. https://www.iconcollective.edu.
6. Levels Music Production. *The 6 Steps to Editing Vocals & Music*. https://www.levelsmusicproduction.com.
7. Masters Traditional Games. *The Rules of Blackjack*. https://www.mastersofgames.com/rules/blackjack-rules.htm.
8. Mel Topf. (2017). *62. Genre, Genre Sets, Genre Systems*.
9. National Low Income Housing Coalition. (2022). Housing needs by State/Colorado. In *National low income housing coalition*. https://nlihc.org/housing-needs-by-state/colorado.
10. Rodriguez, J. (Aug. 2023) *There Are Way Too Many Real-Estate Agents*.
11. Rubino, J., Klamann, S., & Svaldi, A. (2024). *Colorado's affordable housing crisis has spread from the mountains to the Front Range*. https://www.denverpost.com/2024/03/10/affordable-housing-zoning-code-fix-front-range/.

Chapter 7
Application of Systems Principles

Abstract Chapters 1 to 6 give us a good start to learning the systems thinking philosophy that is based on the generic *systems principles*. These are always helpful when they guide our understanding of complex real-world problems and the systems that influence them. However, as we stated earlier, systems thinking is not just about learning fashionable new terminology or appearing to be doing the right thing: it is all about the right ways of balanced *thinking and doing*. So, we disapprove of young fresh talent learning to talk but not to walk! Section 5.1 already listed many common ways of thinking and behaving to avoid. But this is not enough because the question remains: how do we *think, act, and behave*? The purpose of this chapter is to give guidance that is applicable regardless of application or domain. Section 7.2 gives *ways of the systems thinker* (or *SW*s for short). The *SW*s are the starting point for how to apply the *SP*s. The *SW*s are rooted in the *SP*s. They guide our behaviors, decisions, and actions as we try to understand real-world problems or the systems that may influence them, regardless of the exact question, domain, or application. This chapter also introduces some essential tools to kickstart the application of systems thinking.

7.1 Application of Systems Principles

There is no standard way or particular order to apply the *systems principles*. They can be applied one by one or several at once. The *holism principle* is the first stated principle in Section 5.3. Applying this principle would guide us to find and explore the *parts, interconnects,* and the *purpose*. It also guides us to find out in what ways the whole is more than just the parts, interconnects, and purpose. Yet, depending on the problem, it may be beneficial to apply the *3-systems principle* first, which guides us to define the *SOI*, *context systems*, and *enabling systems* first.

After some training and experience in applying the systems principles, they become a more or less natural part of our thinking and approach. With more training and experience, we can apply several principles simultaneously and automatically act and behave in a way aligned with systems thinking. For beginners, however, this chapter suggests two paths that need to be done simultaneously:

K. Eftekhari Shahroudi et al., *Practical Systems Thinking*,
https://doi.org/10.1007/978-3-031-89394-0_7

1. Following an iterative process for direct/literal application of *systems principles*. This process is explained in Section 7.1.1 using the 737 MAX Problem described in Section 3.3.
2. Adhering to simple high-level behavioral guidelines called the *ways of the systems thinker*. This is explained in Section 7.2

7.1.1 Direct Application of Systems Principles

Systems thinking attempts to methodically improve our mental models as we approach complex real-world problems. The first suggested path is to use the *systems principles (SP)* listed in Sections 5.3 to 5.10 directly or literally to formulate the questions we ask to understand the problem better. The recommended path for direct application is an iterative process shown in Fig. 7.1: we first use our intuition or guesswork to pick one of the systems principles that appears to be relevant to the problem and a good starting point. Then, we attempt to improve our understanding of the problem by asking questions about the main features

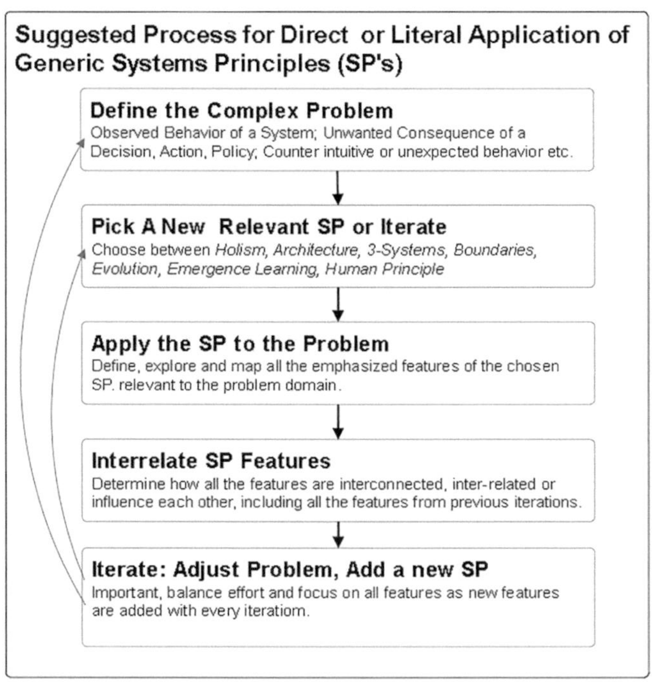

Fig. 7.1 The process of iterative and incremental application of *systems principles*. Observations show that this can become almost second nature with training and experience

emphasized by the *systems principle* we picked. (Recall that the main features of a principle are already emphasized in *italics* font inside the principle statement.) We then ask an essential question of how these futures are interconnected or influence each other. As we ask and answer these questions, we understand the problem better in a more structured way. This better understanding then leads to adjusting the problem definition, picking a new systems principle for the next iteration, and so on.

For example, let's say that we are responsible for delivering the 737 MAX MCAS software that does not have the safety issues discussed in Section 3.3. So, from this perspective, the system of interest is the MCAS software. Let's choose the *3-systems principle* as our starting point. The principle states:

3-systems principle: *systems* come in (at least) *3s*: *context systems*, *systems of interest (SoI)*, and *enabling systems*.

The *3-systems principle* guides us not to limit our efforts and understanding to within the scope of what we are responsible for. It is guiding us to also focus on the other features of the principle: *enabling systems* and *context systems*. The MCAS software is a part of the flight control system (FCS) and is physically loaded into the flight control computer hardware. The FCS is a safety critical part of the airframe or aircraft. So, we now have two context systems that the SOI fits into: FCS and the aircraft. Moving on to *enabling systems*, we see that several sensors are required for the MCAS software to work, including the angle of attack sensors. So, MCAS's sensors are an *enabling system*. Are there other relevant enabling systems? Yes, the proper functioning of the MCAS requires maintenance, so the maintenance system includes the maintenance crew and their equipment. So, maintenance is another enabling system, and so on, we continue to define all the systems that fall into one of the 3-systems categories that are likely to be relevant to the problem.

Next, let's choose the *emergence principle*. This principle states:

Emergence principle: *interactions*(including feedback) within and between the 3-systems cause (sometimes unwanted) *emergent behaviors*.

This principle suggests that interactions within and between the 3-systems may be responsible for the unwanted emergent behavior of the 737 MAX, so this principle encourages us now to put our efforts and focus toward understanding how the 3-systems identified above are interrelated, interact, and influence each other.

Direct or literal application of *systems principles* is the easiest initial step of broadening and deepening our understanding of complex problems or systems that may influence them. But how do we know which principles are relevant to a particular problem? In other words, how do we know which principle to pick? The answer is: SPs are all always relevant and applicable to the problem! The truths contained in all of the principles are ALWAYS helpful! However, depending on the actual problem, our roles, responsibilities, and the person's thinking, one of the principles may provide an easier or more convenient starting point or a more efficient start. However, since these principles always hold, it is more a matter of convenience and preference and not critical, which *SP* we start with.

The key to the iterations is that they are additive: as new *SP*s are picked and their features are added, then the effort and focus are not just purely spent on the features of the new *SP* picked for the next iteration. Instead, the effort and focus remain balanced on all the accumulated features and how they are interconnected or influence each other. We need not continue the iterations until all the *SP*s are exhausted, and we can stop and start it any time necessary.

We observed that most graduate learners did not naturally utilize concepts, features, or truths similar to those contained in the *SP*s when approaching complex or difficult-to-understand real-world problems. Yet, after a semester-long training on their profound meaning and application, they could typically pick one or several *SP*s to structure their thinking and approach to the real-world problem they were interested in. Since our schooling, education, and culture emphasize linear compartmentalized thinking, graduates of both technical and nontechnical disciplines have a more difficult path to systems thinking as they have to change or unlearn mental models that were reinforced and rewarded over a long period. We fully expect that fresh young talent starting college (or even earlier in high school) would have an easier path as they have less to unlearn.

As we apply the iterative process of Fig. 7.1, we deepen and broaden our understanding of the problem. Still, at the same time, we are creating more data, information, charts, and artifacts that are all somehow interconnected. At some point, the volume of material is such that it can no longer be efficiently tracked and executed inside the unaided human brain or even a team. Here is where rigorous systems thinking methods, including various modeling and automation methods, enter the picture. They amplify our capabilities to deal with this situation. For example, DSM (design structure matrix) allows us to visualize many sequenced interconnected blocks' feed-forward and feedback input/output patterns. Causal loop diagrams (CLDs) are a way to understand and communicate complex emergent dynamic behavior of interacting, reinforcing, and balancing loops. Part III of this book is dedicated to teaching the application of rigorous methods (frameworks, tools, and languages) that have proven very useful in expanding our capabilities to deal with real-world problems and systems that influence them.

7.2 Ways of the Systems Thinker

The ways of the systems thinker shown in Fig. 7.2 (or systems ways or SW for short) are four high-level level easy to understand and follow guidelines to ensure our behaviors, decisions, and actions are appropriately aligned with the *systems principles* so that our chances for success improve when we deal with real-world complex problems. It is a way to enforce the "don't just talk, walk the walk" idea. It's also a high-level approximate measure or checklist of whether we exercise systems thinking.

The *SW*s encourage us in four major ways:

1. *Balance* effort and focus on *features* of *systems principles*.
2. *Predict* and *reflect* on the consequences *before* and *after* implementation.
3. Watch for *win/lose* mindsets
4. See yourself as *part of the system*.

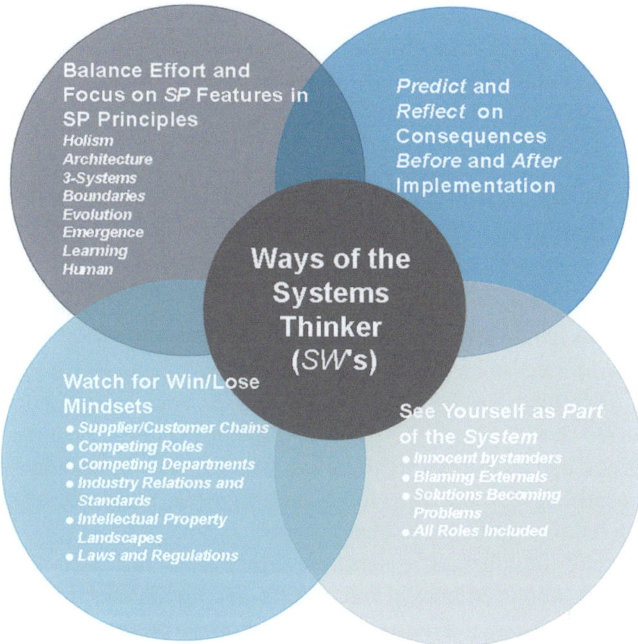

Fig. 7.2 The *ways of the systems thinker* are suggestions to help us align our thinking, approach, decisions, and behaviors with the *systems principles*

Out of the four *SWs*, only the first one is necessary and sufficient because if we genuinely balance our efforts and focus among the interrelated *SP features*, then all suggested behaviors in the next 3 *SWs* will follow naturally. So, in principle, as long as we correctly understand the depth, breadth, and subtleties of each *systems principle*, we only need to ensure our efforts and focus are *reasonably* distributed or balanced among their *features* and how they are interrelated or interconnected or influence each other. Recall that we are not trying to "boil the ocean" or pick all the *SP*s at once, and we are working iteratively and incrementally as shown earlier in Fig. 7.1. For example, let's say we first focus on how and how much *more* the whole is compared to the *sum* of its *parts*. This implies we need to understand the *elements, interconnects,* and *purpose*. If we next define the *3-systems* and how they are interconnected, then at the next iteration, with the incremental understanding we gained in the current iteration, we have to go back and adjust our earlier understanding of how the system is *more* than the *sum of parts*. This way, we ensure our behaviors and actions are balanced and our understanding is not compartmentalized or siloed.

Unfortunately, although we only officially need one *SW*, unanticipated consequences, win/lose mindsets, and blaming external factors are so common in individual, team, department, and organizational behaviors that we felt that they deserved their own separate *SW* so we are not blindsided by behavior that is counter to systems thinking. This is how we arrived at four *SW*s even though just one *SW* is enough. Still, the four *SW*s are still not too many and easy to remember. They are discussed in more detail next.

7.2.1 Balancing *Effort and Focus on* Features *of* Systems Principles

We often hear that "systems thinking" is mainly about thinking out of the box. This statement could not be further from the truth: systems thinking does not favor a particular view like the *outward* view over the *inward* view. But it does encourage us to balance our efforts among the relevant views. So, if *outward* and *inward* views are the only relevant views, then systems thinking encourages us to balance our efforts and focus between them. If there are five relevant perspectives, we should consider them all because the pure focus on a favorite, standard, or specific view has previously led to significant issues.

As we learned in Chapter 5, systems thinking is not just about views or perspectives. It is about balancing our efforts and focus among the *SP*s and all their emphasized *features*: *elements, interconnects, structure, behavior, purpose, life cycle, evolution, 3-systems, interactions (feedback), mental models* and *human reach* altogether need our attention when we are dealing with real-world problems or systems that influence them. So, the first *SW* suggests we should not overfocus our attention and efforts on one perspective or concern.

In practice, even in best-in-class organizations, we see behavioral patterns that are not aligned with systems thinking. Here are a few common scenarios of lack of balance:

- Focusing all effort and attention to what is inside the *SoI*. After all, the *SoI* is what we deliver and are rewarded for. So, why would we not put 100 percent of our focus and efforts on the specific scope of the SoI? The answer is that many unwanted and unanticipated issues emerge from the interaction of the *SoI* with the enabling and *context systems*. So, pure focus inside the SoI boundaries often does not guarantee the success of the SoI if there is some level of complexity or a significant amount of interactions between the 3-systems. A better approach would be to think and understand one or two levels outside the boundaries of the SoI, in addition to obviously spending effort and focus inside the SoI. For example, in the mosquito fighting story in Fig. 5.13, even though our goal is to reduce malaria, we should expand our efforts, focus, and understanding to other species and how their survival might be impacted by our actions to poison the mosquitoes.
- Picking a quick solution versus understanding the need after implementation. This is a common occurrence in a mature industry. The pressure to cut costs and deliver quicker inadvertently often leads to picking a quick solution without considering good alternatives. This can lead to products and services that deliver

the envisaged specifications and pass internal qualification tests but fail to deliver value when implemented and ultimately fail. A lack of balance here can make the "quick" solution very "long and expensive". The Boeing 737 MAX solution did appear to many inside the company as relatively quick, but it ended up costing $50 billion right after entry into service in addition to knock-on effects on other company products that did not even use the same MCAS software,

- Short-term thinking versus considering the long-term consequences of our solutions that may be outside the narrow focus of our goal. One excellent example is the process of lawmaking in the USA. New regulations and laws are passed with a very narrow focus and without a thorough consideration of conflicts and overlaps with other existing laws. This results in lengthy, inefficient, and expensive legal procedures, approvals, and conflicts.

7.2.2 Predict *and* Reflect *on the Consequences* Before *and* After *Implementation*

Not all problems or consequences of our behaviors, decisions, and actions are predictable. Yet, there are so many examples, such as the LED icing example, where it is easy to see that with some reasonable effort (such as simple communication between the LED expert and the headlamp lens expert), the icing issue would have likely been predicted and resolved before service entry. Yet, several automotive manufacturers with decades of pedigree were surprised as they started incorporating LEDs into their headlamps.

When it is impossible to predict unwanted emergence, we need to review the consequences retrospectively after implementation. Yet, in many cases, such as laws and regulations in the USA, we are not even looking for any data or information to help us reflect on the impact of a new law or regulation even after implementation! So, we have a vast body of laws and regulations without a clear understanding of their consequences, interconnections, overlaps, and conflicts.

Balancing our efforts and focusing on all the SP features will avoid this problem. Still, this issue of not worrying about the unanticipated consequences of our solutions (before and after implementation) is so common that we decided to make it a specific *SW* to cover it:

Predict and *reflect* on the consequences *before* and *after* implementation.

7.2.3 Watch for Win/Lose *Mindsets*

Our dominant mental models are formed by our culture, education, experience, and evolutionary effects. For example, a survival mode model can turn on, sometimes without us being conscious of it, when we face people, situations, or factors that are

Fig. 7.3 4X400 m relay requires the best individual running and the best baton passing. Competing while collaborating is required in many real-world situations; however, our win/lose mindset can inadvertently turn on and dominate where collaboration and a win/win mindset are more appropriate

new, strange, or unfamiliar. For example, if our ancestors met a lion in the forest, the ones that did not trust it and ran away survived and procreated so that we now have a fight or flight mode of operation that can turn on sometimes appropriately and sometimes inappropriately.

The win/lose or zero-sum mindset is a dominant mental model related to the scarcity of resources and survival of the fittest in our evolutionary path. In many cases, this mindset is appropriate. For example, competing in the 100 m individual sprint in the Olympics requires this mindset. However, issues arise when this mindset dominates in situations where it is not appropriate: in a 400X4 relay race, you not only have to win your segment, but you also have to pass the baton and set your teammate up for success as shown in Fig. 7.3. In this case, you must have a win/win mental model when dealing with your teammates while having a win/lose mindset against competing teams. Under pressure, it is so hard to balance these, and we have to be careful not to let the win/lose turn on and dominate automatically where it is of no benefit or when it works against us. At the 2024 Paris Olympics, two high jumpers were offered gold after their score tied. They refused despite a precedent at a previous Olympics where two high jumpers received gold. Their win/lose mindset was so strong that they continued to compete until one lost, so there was only one gold. We are not saying the athletes' decision was wrong; we are only using this story to illustrate how dominant a win/lose mindset can be to the point of risking a gold medal already in hand!

While a win/lose mindset may be appropriate in some situations, there are unfortunately too many real-world situations where people on the same team enter a win/lose mindset and forget that they can only win if the entire team, department, organization, or chain of organizations can win together. This can be amplified by the inadvertent side effect of inadequate reward structures, ways of measuring performance, or ways of defining our roles and responsibilities. Other factors such as untimely, inadvertent, or inappropriate turning on win/lose mental models also play a role. Several real-world scenarios are described in the following subsections to help improve our abilities to identify win/lose mindsets that are not appropriate for approaching the complex real-world problem at hand. In these scenarios, the advantages of an approach, action, or decision only benefit those adopting a win/lose mindset in the near term. At the same time, the disadvantages only impact another person, team, department, or organization. A typical outcome is that, eventually, all lose. The following examples are true, common real-life stories where the real story has been simplified and shortened for brevity. Furthermore, identifying information has been omitted for obvious reasons:

Role Conflict Scenario: Systems Engineers Versus Project/Program Managers

In many mature industries, the responsibility for understanding and delivering what a product or service shall do is assigned to a systems engineer (SE). Then, the responsibility for delivering the product on cost and schedule is typically assigned to a program or project manager (PM). Often, the performance and rewards for these two roles will be compartmentalized, meaning the performance of the SE and the PM are measured by what they are directly responsible for but not by a joint measure. The SE's concerns about understanding the customer and application needs and requirements may take longer, while the PM wants to pick a solution and run with it due to schedule and cost pressures. As pressures mount on both SEs and PEs to meet their individual goals, one or both can forget that they only win if they both win. If the leadership at the top of these two functions is shortsighted, they may give short-term priority to the PM (only short-term priority). Or they may give priority to the SE (only long-term priority). Favoring one over the other is, in many cases, a disaster: in one case, we risk having a product that does not do what it needs to do in the real world to succeed. On the other hand, the product will be costly and late to market. We have known of actual conflicts between these two functions where the conflict caused one or the other chief to be fired! There have been so many repeated patterns of SE/PM conflict in the industry that INCOSE has published a textbook on the topic of integration of SE and PM disciplines and has an active working group to create work products that companies can use to smooth over the collaboration between the SE and PM roles.

Finger Pointing Scenario: Blaming Top Contributors for Failures

A systems thinking and modeling team is created in a reputable company delivering safety-critical products. This is achieved by putting together several very collaborative specialist A-players whose task is to solve multidisciplinary complex systems challenges to all other departments that face problems they cannot solve. Soon, this team became very popular within the business unit, and they built a multiyear outstanding track record, receiving raises, promotions, and customer praise. However, creating this team was a very rare, unprecedented initiative as A-players are sought after in every department, and it was a tough decision to move them out of their regular departments to work on systemic challenges in a team.

This worked well for several years until a new director was hired externally, with previous executive leadership experience at the vice presidential level in a reputable international company. To show short-term progress, the director started to hide key information created by members of this A-team from the customers who questioned the reliability and life of a brand-new product that promised to be very lucrative.

Members of the A-team who resisted this were sidelined, excluded from major customer meetings, and subject to extra scrutiny and blame. Soon, reliability and life issues started to surface during customer tests. In some cases, the new director did not renew the A-players' contracts, and his toxic win/lose mindset caused a few others to escape by moving out of the department. All this happened while the top leadership was busy with other business challenges and believed in and supported the director's reasoning. Soon, because of continuous critical failures of the new product in customer tests and information that surfaced belatedly about fundamental problems with the design of the new product (that was correctly flagged by the original A-team players a couple of years earlier!), the director was fired.

The product was canceled, causing the company a significant loss of opportunity. So not only did the director adopting the win/lose mindset lose, but the A-team suffered a significant disruption, and the company suffered a loss. The real story has many interesting twists and turns pointing to a lack of systems thinking: a powerful individual imposing his severe win/lose mindset. At the same time, the organization was distracted and incapable of detecting and correcting the toxic environment that his lack of systems thinking caused, leading to a loss for everyone involved.

Blocking Progress Scenario: Killing Initiatives That Help!

A motivated and experienced employee notices that many new technology proposals fall into the "Valley of Death (VoD)". This is a well-known term for common phenomena in industry that refers to proposals that absorb a lot of funding but never make it into a final product that enters into service. The employee makes it his mission to try to understand the "Valley of Death" by comparing the characteristics of several previous proposals, some of which successfully entered service while others fell into the VoD. He got permission from his leaders to work on this. He used a rigorous systems thinking method to predict the probability that a new proposal might fall into the VoD before any expenditure occurs. He was excited about his progress because if his method worked, his predictions would reduce waste while giving feedback on improving the proposals so their chances of escaping the "VoD" would improve. Saving dollars and improving outcomes was a fascinating potential indeed.

The employee demonstrated his initial results to his organization, who was impressed with his progress. He then asked to get access to a more extensive database of proposals and projects to fine-tune his method and validate it. This time, one of the senior leaders rejected his request and asked him to stop and abandon his project! When promising new efforts that are well aligned with the organization's stated goals are thwarted, the cause may be individuals whose win/lose mindset has turned on dominating over efforts, thinking, and seeing win/win opportunities. The senior leader explained that he identified a potential unwanted side effect of the VoD method. Other than helping fewer proposals fall into the VoD, it could also be used to identify ongoing favorite projects that have a significant probability of falling into the VoD, causing cancellation of ongoing projects.

We are not claiming that the senior leader was wrong in identifying an unintended consequence of applying the VoD tool. On the contrary, we know that as systems thinkers, we should always try to anticipate unwanted consequences of implementing our solutions. So, the leader identified a potential side effect of implementing a new solution. What the senior leader did wrong was adopting a win/lose mindset in favor of continuing ongoing projects at the cost of improving the organizational capability to deliver a bigger bang for the buck for all future projects. Instead, the leader should have adopted a win/win mindset, i.e., find ways to mitigate the potential unwanted side effects to ongoing projects while removing the roadblocks to apply the new VoD tool to improve the future success of his organization, which is also part of his responsibility.

7.2.4 *See Yourself as* Part of the System

Have you noticed that as we enter late summer and fall in the USA, city, state, and federal budgeted infrastructure projects become a massive nuisance to the larger population because of causing significant traffic delays? Why are these projects not as visible or annoying during the other times of the year? The simple answer is the *use it or lose it* regulation regarding local, state, and federal budgets. In other words, as we approach the fiscal year-end, where the budget closes, there is a concern that any unspent amounts will be lost and not included in the next budget year or cycle. The logic is that unspent amounts were unnecessary and, therefore, a candidate for being dropped during the next budget cycle.

One way around it might be to spend the money as fast as possible, as soon as the budget year opens! Well, this is not practical for several reasons. One reason is that the entity managing the expenditure worries about running out of money to cover unforeseen circumstances such as natural disasters, strikes, etc. The other reason is that infrastructure projects, like many other categories of projects, experience cost and schedule overruns. Issues like these add up to underspending during the early months of the year, followed by turning on all possible projects as we get closer to the budget year-end.

Another example is the amount of magnesium chloride used by city, county, and state services in areas that suffer harsh winters, like Colorado and Wyoming. They stockpile large amounts of this salt to have enough to melt snow/ice and clear roads in winter storm conditions. But some people observed these services spreading enormous amounts of this salt at the tail end of the cold season when the weather was not so severe! Magnesium chloride damages road surfaces and encourages corrosion of the car body and chassis in addition to building up in soil and water. It is toxic to the environment and the ecosystem, including trees, animals, and humans. Magnesium chloride is an FDA-regulated drug, and significant intake of it is known to cause respiratory depression, hypothermia, flushing, low blood pressure, stupor, and sweating. So, why would a city spread more than necessary magnesium chloride on the roads? The likely reason is probably the *use it or lose it* policy that any unused amount of the magnesium chloride stockpile will not be replaced next year.

Another example of not seeing yourself as *part of the system* is the typical pattern of blaming "external factors beyond our control" or pointing fingers at others or natural disasters such as pandemics to avert blame! This dominant self-serving mental model prioritizes blaming anything or anyone for the issues versus understanding how we may have played a role or caused the issues we face. Figure 7.4 shows a reinforcing pattern of behavior exhibited by US federal debt for nearly a century. This number does not include many other forms of debt, such as local state, city, and personal debt, but its exponential growth for decades is alarming. According to the US Treasury, the US debt is already larger than the US GDP: *"The average GDP for the fiscal year 2023 was* \$26.97 *T, which was less than the U.S. debt of* \$33.17 *T. This resulted in a Debt to GDP Ratio of 123 percent"*. Yet we rarely see politicians or lawmakers from either party display signs indicating they might be aware of their role in creating the gigantic problem.

From the perspective of the budget controller rewarded for optimizing this and next year's budget, or the person in charge of the decision to overuse magnesium chloride, or the politician pushing to increase US debt, all is well because their mental model does not allow them to see themselves as part of the system or care about issues they perceive to be outside their immediate area of responsibility. Their mental model will not detect any problems they cause that fall outside their immediate short-term focus. However, the people, plants, animals, and businesses going about their activities are all adversely impacted! When we do not see ourselves as *part of the system*, we become self-centered or internally short term focused on the system of interest so that we do not worry about the impact our mindsets, actions, decisions, solutions, and behaviors have on other people, businesses, society, and the environment.

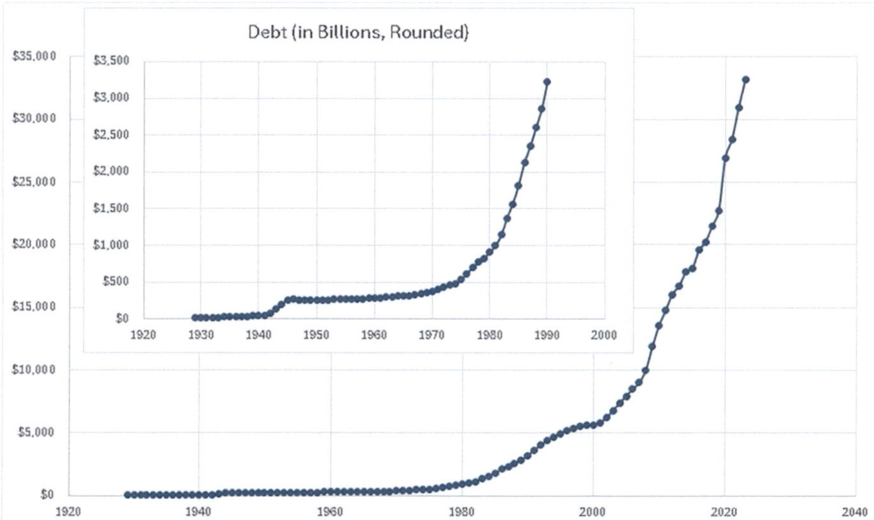

Fig. 7.4 The US debt has been exhibiting reinforcing behavior for nearly a century. Yet, the dominant political parties and members of the Congress have largely blamed others or external factors beyond their control. Source: Fiscal Data, US Department of the Treasury. "Historical Debt Outstanding"

As mentioned earlier, balancing effort and focus on the features of systems principles will naturally drive the correct behavior. However, not seeing yourself as *part of the system* is so common that we felt it necessary to include it in *ways of the systems thinker*. On the other hand, those who do see themselves as part of the system achieve more sustainable results for themselves and others. For example, scientists utilized systems thinking when they collaborated internationally to create a COVID-19 emergency use vaccine in a record time. Figure 7.5 shows the large number of countries that received a significant vaccine donation from the USA (more than 690 million doses), which is a good example of not limiting our efforts and focus inside the immediate area of our responsibility.

At this point, we must acknowledge that seeing yourself as part of the system can be challenging. It requires *personal mastery* and an *organizational culture* that fosters systems thinking. For example, how would we, as systems thinkers trying to see ourselves as part of the system, take action when the rest of the system, especially the incentives at play, are arranged against us? Chapter 9 gives guidance on how to gain personal mastery, how to use the tools for bottom-up learning, and how to improve the culture of the organization to foster a culture of systems thinking based on what we have practiced and experienced in our organizations and projects.

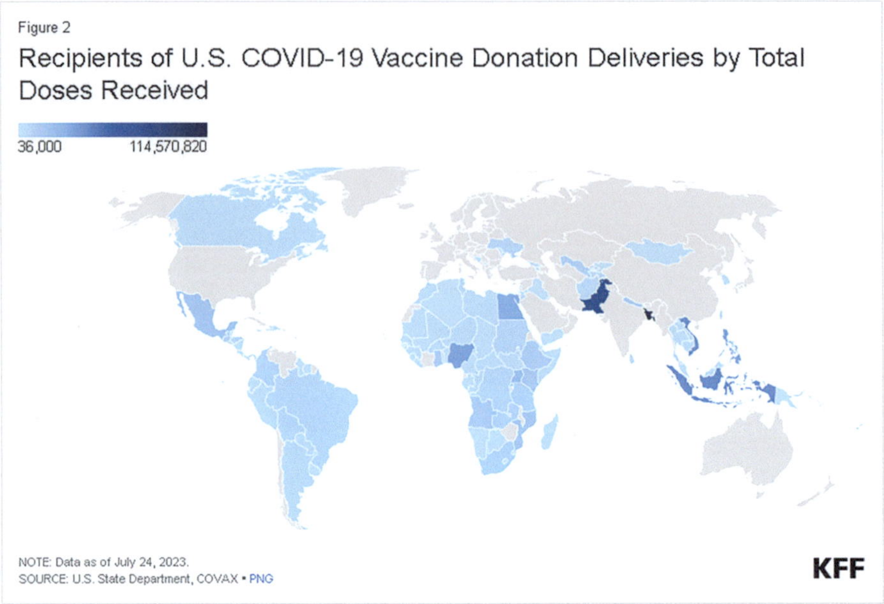

Fig. 7.5 The USA donated circa 700 million doses of COVID-19 vaccine internationally, showing a mental model that sees itself as *part of the system* (Source KFF website)

7.3 Using Rigorous Methods to Guide Our Actions and Behaviors

The direct application of *SP*s using only our heads is a great start but creates a relatively large amount of data, behavior patterns, information, charts, and artifacts. Soon, this outgrows our ability to track, manage, and understand. We, therefore, need reliable methods to help individuals, teams, and organizations. By methods, we mean frameworks, languages, and tools that expand our capability to apply the principles of the system and exercise systems thinking.

This chapter briefly introduces some methods, tools, or artifacts that have proven helpful in many technical and nontechnical disciplines. These are:

- System maps and mind maps: Used by systems thinking rigorous method called architectural modeling covered in detail in Chapter 12.1.
- System architectural modeling uses rigorous tools and formal modeling languages to create digital representations of a system of interest and its environment. Architectural modeling allows for system visualization, including structural and behavioral modeling. Both a system's physical structure and a functional representation can be illustrated. (See Part III chapters that review model-based rigorous tools for systems thinking).

- Commonly used architectural models for systems thinking include:

  Context diagrams: Present a structural view of a system of interest- some looking inward at subsystems and components. They also can model outward, illustrating the 3-systems principle of every system operating in its environment with other systems.
  Use case diagrams: These capture the key functionality of a system modeling the system actors, both operators and other systems interacting directly with the key functionality.
  Activity diagrams: These are functional models illustrating a sequence of actions that occur to fulfill intended functionality. These are useful for a high-level overview of functions for a given system use case. They assist in systems thinking by modeling condition-based relationships and activities flow within a system's intended behavior.

• Causal Loop Diagrams (CLDs) and Stock & Flow Diagrams: Used by a rigorous systems thinking method called *system dynamics* covered in detail in Chapter 12.

Next, we learn some basic rigorous methods like *system mapping* and *causal loop diagrams* to visualize a system. Then, we review a range of real-world applications in Chapter 10 to show how rigorous methods are applied to understanding real-world problems.

7.3.1 Starting with a Blank Sheet

For many early systems thinkers, the most challenging aspect of moving from a thought exercise to applying rigor is staring at a blank sheet of paper, wondering how to begin to capture the ideas on behavior, information, and connections of complex systems. If systems are interconnected, where should we begin? Is one element of the system more crucial than others? We can tackle this challenge and build momentum by knowing that, yes, systems are interconnected and that we can consider that the system's behavior will identify crucial elements of the system. These considerations allow us to start anywhere in our thoughts and let the system structure evolve. One way to start is through system visualization.

7.3.2 Visualizing the System

An early starting point is visualizing the system through diagramming tools like mind mapping, system maps, or context diagrams. Each tool helps us document our understanding of a complex system. Visualization also differentiates systems thinkers from linear thinkers. Take the iconic neighborhood lemonade stand shown in Fig. 7.6. A child is gleefully offering a refreshing drink to a potential customer,

Fig. 7.6 Can we think of a neighborhood lemonade stand as a system?

and other potential customers are seen in the background. At first glance, we might only notice the relationships between the customer and child, but the *3-systems principle* states that other connected systems exist. Visualizing a system like a lemonade stand allows us to uncover these connections, helping us consider resource management needs, optimize operations, or understand interactions that influence outcomes like profitability or customer satisfaction. While a lemonade stand is a system that many children and parents successfully build without much consideration, we present it here as a simplified example of using visual mapping. With visual mapping, we can start asking questions like how many lemonades can be sold. We can also pose potential problems that need to be addressed proactively. We might see the bucket of lemonade is empty, so we should buy some lemonade before it runs out. We see a lot of cups for serving lemonade, so perhaps they aren't running out but have more hidden away. We can conclude that the lemonade is fresh if we see fresh lemons at the lemonade stand. We can also mentally list elements of the system: lemons, cups, a table, a child, a sign with the price, pictures, and a change box. What is more difficult is to draw connections between these elements mentally. To help us visualize these connections, we can use system mapping and mind mapping.

Mind Map

Mind Maps Mind mapping is a process for mapping the whole system and how its parts are connected. Mind mapping allows us to visualize the relationships between different system elements, making identifying patterns and potential leverage points easier. A mind map is a holistic process that maps the system by creating a hierarchy of interconnections. Often, mind maps can be used for strategic planning, brainstorming sessions, and team collaborations.

Let's explore mind mapping our neighborhood lemonade stand. We first start by defining a central concept, which is our lemonade stand. We begin by placing this concept at the center of our workspace, such as Fig. 7.7.

Fig. 7.7 The beginning of a mind map for a lemonade stand

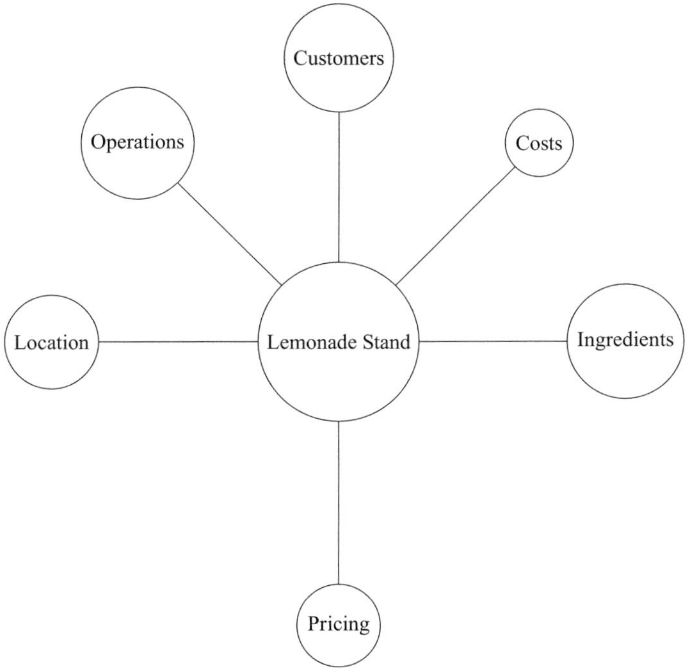

Fig. 7.8 The second layer for lemonade stand mind map

From this central node, we can identify categories of thought that relate to the lemonade stand. These categories will serve as the main branches in our map. For our lemonade stand, we can consider items such as: **ingredients, location, customers, pricing, costs, and operations**. We add these categories around our central node, creating a hierarchy shown in Fig. 7.8.

From this secondary node, a mind map will visualize elements that branch out, representing additional themes. For instance, what ingredients are required for lemonade: ice, sugar, lemons, and water? What are some ways we might price the lemonade: per cup, per different sizes of cups, the same as other lemonade stands, or lower? Who are our possible customers: kids, parents, or passersby? What is the best location: near a park, a school, or in front of the house. From each secondary branch, a map will grow, representing additional concepts until our brainstorming is complete. Many mind maps are also color coded to visually represent a system's layers. Figure 7.9 illustrates how the hierarchical structure of a lemonade stand is visualized.

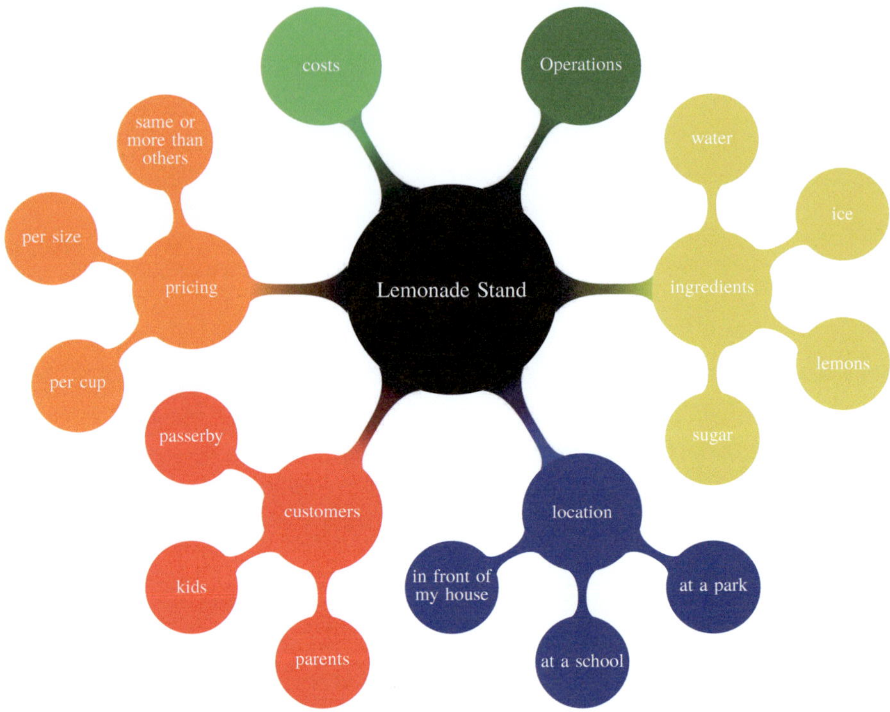

Fig. 7.9 An evolving mind map of a lemonade stand

Systems Maps

Mind mapping is a valuable tool for brainstorming components of a system. Still, to explore and contextualize system interactions, system maps provide a more powerful tool for visualizing the relationships, feedback loops, and interactions in a complex system. System maps represent and understand the interactions and dependencies within a system. Systems thinkers use these maps to show how different parts of the systems affect each other in a simple form. To further visualize our lemonade stand, let us start with the same core components we identified in our mind map and consider how these items function:

1. **Lemonade stand**: The physical structure where sales take place
2. **Lemonade ingredients**: What is being offered for sale
3. **Pricing**: The cost of lemonade
4. **Location**: The place where the lemonade stand is located and a sale occurs
5. **Customers**: The individual or individuals that purchase lemonade
6. **Costs**: The money that is spent to be able to sell lemonade
7. **Operations**: How the lemonade stand will be operated (e.g., alone, with other children, only during the day)

Each of these above characteristics describes the lemonade stand, but we should also add external factors that influence the system and may not be under the control of our lemonade stand:

1. **Weather**: A variable that influences customers and our choice of location
2. **Demand**: A consideration representing how many sales we expect
3. **Net profit**: Our money earned from selling lemonade after subtracting our costs
4. **Competition**: Representing other lemonade stands nearby

With our expanded list, we now define possible interactions between these components and map out the relationships, for example:

1. **Weather** $\rightarrow$ **Demand**: Pleasant weather can bring in more customers and traffic near our lemonade stand, and hot weather can increase the demand for lemonade.
2. **Demand** $\rightarrow$ **Net Profit**: Increased customers mean more people want lemonade, which means more net profit. Decreased demand reduces our profit because we sell less lemonade.
3. **Operations** $\leftrightarrow$ **Lemonade stand**: We will want to consider that if we want two children to sell lemonade, we will need a larger lemonade stand, and likewise if we are limited to a small lemonade stand, we may not be able to have two children at the stand.
4. **Pricing** $\leftrightarrow$ **Customers**: If we price the lemonade too high, we may not attract all types of customers to our lemonade stand, for instance, children.
5. **Pricing** $\leftrightarrow$ **Demand**: Likewise, if we price lemonade correctly, we will attract more customers; if we price it too low, we might sell it too quickly.
6. **Lemonade Ingredients** $\leftrightarrow$ **Costs**: Our lemonade stand depends on making lemonade, and that depends on ingredients, which can increase or decrease our costs.
7. **Costs** $\rightarrow$ **Pricing**: Our costs may cause us to increase the price of our lemonade, affecting our customers (see interaction *g*) and demand (see interaction *e*).

With system mapping, we do not need to list all interactions initially. We can start with those that come to mind easily and add to our lists as you answer the question: how does each influence one other element in the system? A simple rule of thumb is that we should start by listing N*2 interactions. So, for our list of ten lemonade stand elements, we can start by finding 10*2 or 20 interactions. Our next step is to begin to draw these interactions.

In a system map, each component is represented as a node, and the relationships between elements are shown as arrows representing flows. Unlike a mind map, you do not want to draw in terms of hierarchy but in terms of interactions. We can start with a part of our system with many interactions, such as pricing show in Fig. 7.10, and draw these first.

We then continue adding all interactions until we have a final system map, Fig. 7.11, that shows us how weather influences lemonade demand, which affects our net profit. It further shows how location can affect customers, which can also affect lemonade demand.

Fig. 7.10 The start of a system map for a lemonade stand

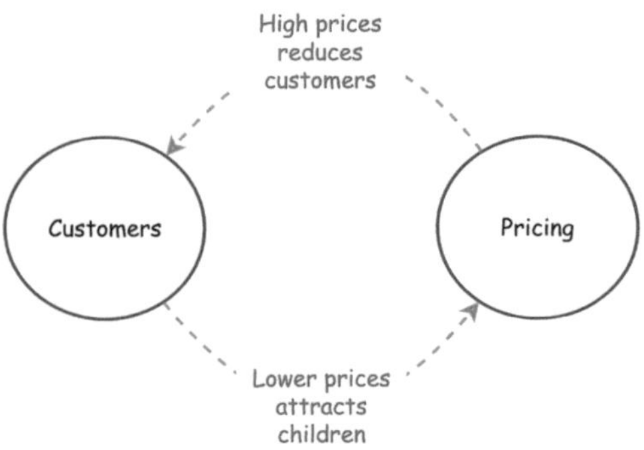

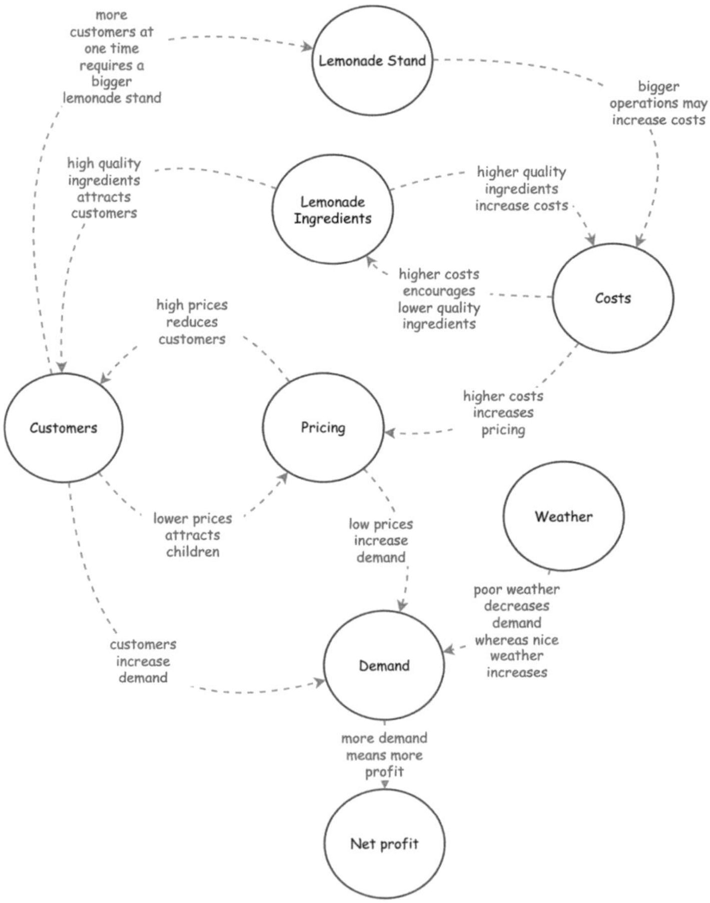

Fig. 7.11 A flushed out mental system map of a lemonade stand

With a system map, we can mentally work through scenarios that affect the behavior of our system. For example, if we move the location of the lemonade stand, will we make more net profit? Or if we increase the price, will we still have enough sales to sell all of our products? What will happen if we do not buy enough ingredients?

A system map, like many maps, allows us to work through different ways we can achieve our goals or look out for potential problems. Building on this visual framework, causal loop diagramming adds the consideration of dynamic relationships and behavior among the system components. It focuses on identifying the feedback loops and potential leverage points. Moving from system mapping to casual loop diagramming focuses on the underlying structure that drives system behavior.

7.3.3 Causal Loop Diagrams

The TEDx ST101 video in Fig. 7.12 is a good initial introduction to how causal loop diagrams (CLDs) help us capture and explain complex emergent behavior resulting from circular (i.e., nonlinear) cause and effect relationships. CLDs help us understand how feedback loops interact to explain the complex nonlinear patterns of behavior we see in complex problems or systems that influence them. The CLD in Fig. 7.12 shows the *fast fashion* CLD with two significant loops. The left loop shows the reinforcing effect of production, inventory, and advertising, leading to more customers and profits that, in turn, feedback to amplify customers. The right loop shows how wear/tear/going out of style also reinforces or amplifies

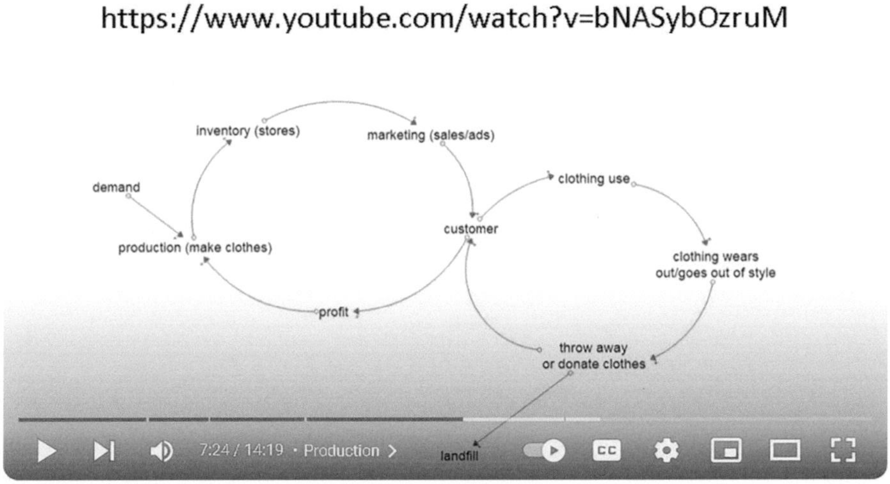

Fig. 7.12 TEDx ST101 YouTube video where Anna Justice explains how CLD helps understand the complex behavior of *fast fashion* and *industrial agriculture*

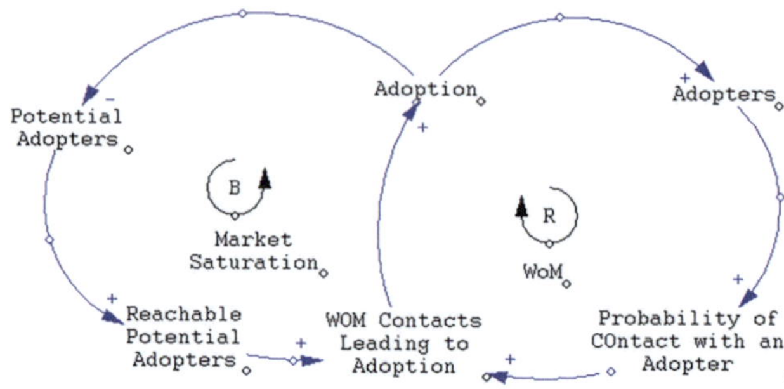

Fig. 7.13 A causal loop diagram (CLD) showing the interplay of the word of mouth (WOM) reinforcing loop with the market saturation balancing loop. The WOM loop amplifies a change in the adoption rate, while market saturation fights a change in the adoption rate. The interplay of the WoM and market saturation loops determine the shape of the behavior of adoption over time

customers. Since both loops reinforce customers and profits, manufacturers have two paths to influence customers and profits: through production/marketing as well as durability of fabrics and longevity of styles. This effect is called *fast fashion* since two reinforcing loops encourage or amplify customers. By itself, it captures the reinforcing effect that profits drive production. To represent complex behaviors, we also typically need balancing loops or loops that try to fight change, contrary to reinforcing loops that try to amplify change. We may also need more than just two loops (i.e., several reinforcing and balancing loops) to represent or model a complex pattern of behavior. CLDs are also a great visualization and communication method between individuals and teams. Figure 7.13 shows another CLD intended to represent the adoption behavior of a new product or technology.

In a CLD, each arrow represents an individual cause and effect and has a "+/−" or "S/O" sign to indicate whether the effect changes in the "same/opposite" direction as the cause. If the behavior of a loop is such that it will try to further amplify a change in a loop variable in the same direction, then we call it a *reinforcing or R* loop. If the behavior of a loop or feedback is such that it will try to oppose or fight a change in a loop variable, then we call it a *balancing or B* loop. Both R and B loops are feedback loops because they imply that a change in a variable of the loop will go around the loop and come back to have the same or opposite effect on the same variable. Figure 7.13 shows that the interplay of the WoM reinforcing loop with the market saturation balancing loop determines the adoption behavior.

7.3.4 Adding Stocks and Flows to Causal Loop Diagrams

Causal loop diagrams are an excellent visualization and communication tool because they attempt to reveal and capture the reinforcing and balancing loop structure that might potentially explain the real-world patterns of behavior we are trying to understand. However, CLDs alone cannot create a simulation or running model that behaves like those patterns. Stocks and flows allow us to go one step further and turn the visualization into a model that integrates the dynamic interaction of reinforcing and balancing causal loops. Stocks and flow diagrams help:

- Add real effects that occur in nature, such as accumulation, delay, inertia, and memory effects
- Replace less intuitive math with the more intuitive bathtub analogy so we can more directly understand dynamic interactions
- Create a running model or simulation that can match the actual pattern of behavior
- Use of running model for scenario (what-if) and sensitivity analysis
- Add policy or control loops to the model to evaluate their effectiveness and anticipate their unwanted side effects

Stocks and flows help us add real effects to CLDs, such as accumulation, delay, inertia, memory, etc. Their addition makes modeling the pattern of behavior more realistic and enables the simulation of complex behavior on a computer.

Figure 7.14 shows what we mean by stocks and flows. Stocks are accumulative, like the level of water in the bathtub. Flows are flow rates into or out of stocks like valves or drains. Figure 7.15 shows the resulting model after we added stocks and flows to the adoption CLD of Fig. 7.13 to make it more realistic.

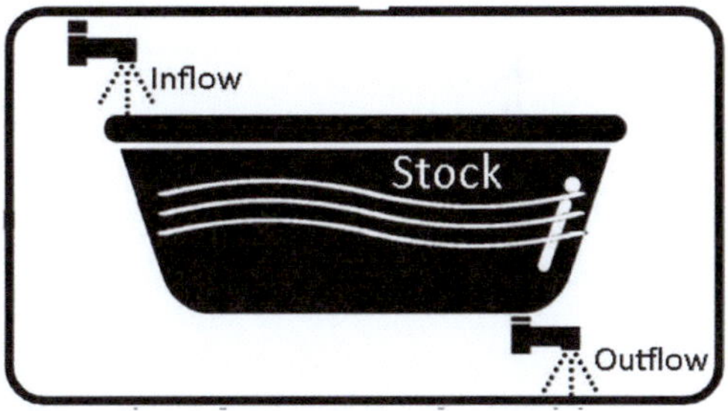

Fig. 7.14 Stocks are accumulative like the water level in a bathtub that accumulates water flow into and out of it (outflow). Flows are like valves or drains that control the rate of addition (inflow) or water removal (outflow) from the bathtub

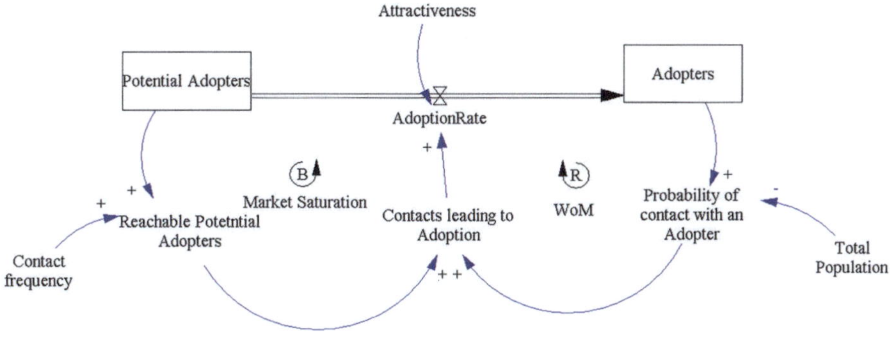

Fig. 7.15 Adding stocks and flows to the adoption CLD makes the adoption model more realistic and allows it to be simulated to enable, for example, sensitivity and scenario analysis

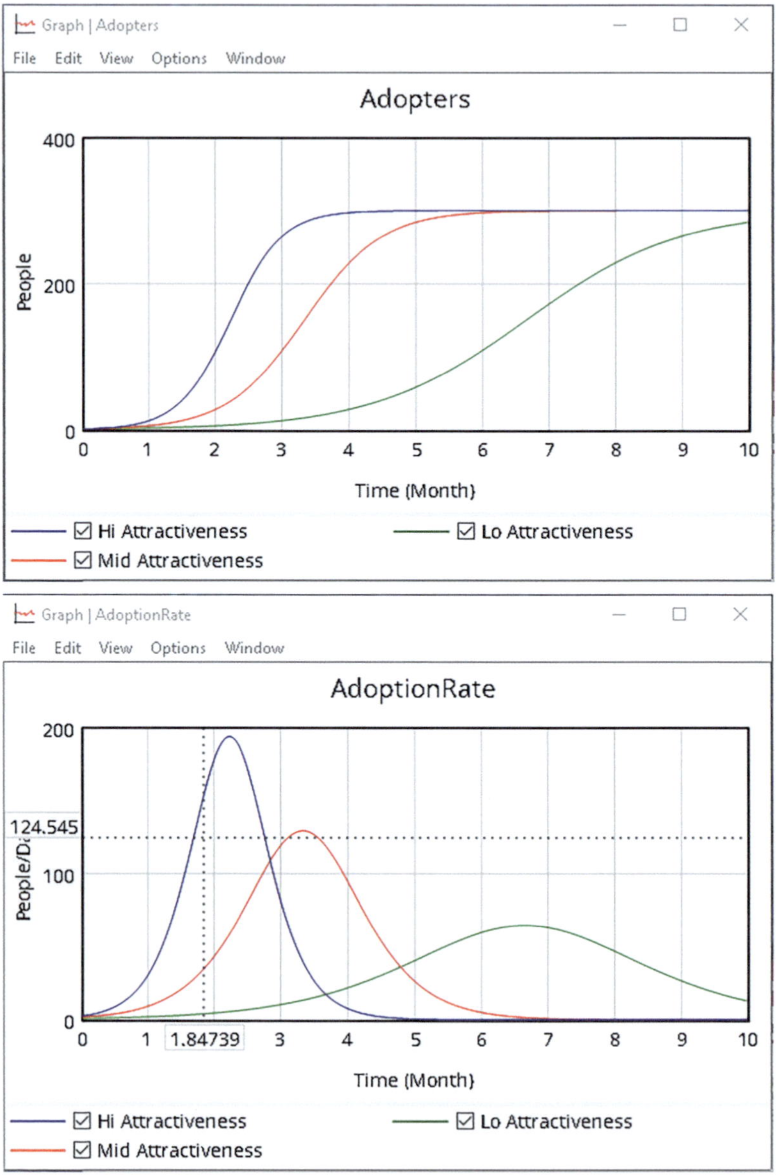

Fig. 7.16 Simulation of multiple attractiveness scenarios for adoption using Vensim software

Now, suppose we know the details of the cause-and-effect relationships corresponding to every arrow, such as equations or lookup tables. In that case, the stock and flow diagram of Fig. 7.15 is simulatable. For example, a system dynamics software called Vensim was used to create the different attractiveness scenario behaviors shown in Fig. 7.16.

7.3.5 Readings

Systems thinking has a long and interesting history. You can understand this chapter without reading beyond this book, but if we are still curious, here are some useful readings:

Donella Meadows, "Thinking in Systems: A Primer" [4], is perhaps the first book (after this one, of course) read by the aspiring systems thinker. She was a prolific author and deep thinker: all her writings are worth reading as you move along your systems thinking path.

Peter Senge, "The Fifth Discipline"[6], provides much wisdom and guidance on applying systems thinking, especially in the social arena.

Kathleen Clark, in "Using Systems Thinking to Shift Mindsets: A Fortune 500 Company's Multi-year Journey" [3], provides a case history of applying ST in a large, real-world corporate setting.

Taleb, in "The Black Swan" [7], explains with many practical and theoretical examples how systems can seem to be executed smoothly until a black-swan event changes everything. The approach is substantially stochastic, though the systems thinker can learn to be humble in recognition that if something can happen, it might, and more often than you imagine.

Chris Argyris, "Double-loop thinking: Double Loop Learning in Organizations," Harvard Business Review, 9/1977, describes an excellent application of systems thinking that provides a practical method of continuously improving oneself or a team's performance and an anti-pattern that suggests the presence of suboptimal performance.

For a deep dive into the mind map, please take a look at the PhD dissertation of Glennis Edge Cunningham's Ph.D. dissertation, "Mindmapping: Its Effects on Student Achievement in High School Biology." PhD thesis, The University of Texas at Austin, 2005. UMI number 3215351. Available from ProQuest.

For an intense dive into the various subcultures of systems thinking, loop awareness, feedback, and more, see GP Richardson, "Feedback Thought in Social Science and Systems Theory" [5].

One of the original proponents of win-win thinking was Barry Boehm. See [1].

The foundations of mental models are established in [2].

Many of the terms used in this book have definitions in the INCOSE Systems Engineering Handbook[8]. This is a good reference developed by an industry/academia consortium for those needing to learn standard terminology, concepts, and processes that apply systems principles to understand and create products and services.

Finally, if we want to continue our ST and SD journey and explore some of its philosophical foundations, we can read publications by Jay Forrester, GP Richardson, John Sterman, and Jorgen Randers.

7.4 Reflection on Chapter 7 Material

In this chapter, we learned that systems thinking is not just about learning fashionable new terminology or appearing to be doing the right thing: it is all about the right ways of balancing *thinking and doing*. We already listed the misconceptions that we may carry associated with linearized compartmentalized thinking in Section 5.1. We then learned about the *systems principles* (*SP*s) in Sections 5.3 to 5.10. These are the pillars of systems thinking and guide our thinking and actions toward a systemic understanding of real-world problems that are otherwise harder to grasp or deal with.

Since the *SP*s are all always relevant and applicable everywhere, and since each has several features, then it becomes a daunting task to simultaneously consider everything all at once. Hence, we created an incremental iterative approach to direct application of *SP*s in Fig. 7.1.

Still, the principles by themselves are insufficient because the question remains: how do we practically implement these principles into our approach and behavior? The answer is the *ways of the systems thinker* (*SW*s). *SW*s teach us to balance our efforts and focus among the *features* of the *SP*s while reflecting on *consequences*, watching for *win/lose* mindsets and seeing ourselves as *part* of the system as summarized in Fig. 7.2.

The rest of Part II of this book deals with identifying and dealing with hurdles in our path toward systems thinking: from the shortage of available systems-oriented skills in Chapter 8 to gaining personal mastery and creating a culture that fosters systems thinking in Chapter 9.

References

1. Boehm, B. W., & Ross, R. (1989). Theory-W software project management principles and examples. *IEEE Transactions on Software Engineering, 15*(7), 902–916. ISSN: 0098-5589.
2. Gentner, D., & Stevens, A. L. (1983). *Mental models*. Cognitive science (viii, 348 p.). Erlbaum. ISBN: 0898592429.
3. Goodman, M., Soderquist, C., & Clark, K. (2015). *Using Systems Thinking to Shift Mindsets A Fortune 500 Company's Multi-year Journey*. Conference Paper. https://proceedings.systemdynamics.org/2015/proceed/papers/P1104.pdf.
4. Meadows, D. H. (2008). *Thinking in systems: A primer*. Chelsea Green.
5. Richardson, G. P. (1991). *Feedback thought in social science and systems theory*. University of Pennsylvania. ISBN: 0812213327.
6. Senge, P. (2006). *The fifth discipline: The art and practice of the learning organization*. Random House Books.
7. taleb2010 Taleb, N. N. (2010). *The black swan: The impact of the highly improbable* (2nd ed., xxxiii, 444 p.). Random House Trade Paperbacks. ISBN: 081297381X 9780812973815 (pbk.).
8. Walden, D. D., et al. (2015). *Systems engineering handbook: A guide for system life cycle processes and activities* (4th ed.). Ebook Central (EBC). Wiley. ISBN: 9781118999417.

Chapter 8
Shortage of Critical Systems-Oriented Skills

Abstract Systems thinking skills required for understanding real-world complex problems are, at the time of writing this book, in high demand for specific complex systems-oriented jobs dealing with hardware, software, cyber-physical, data, autonomous, social, production, maintenance, operations, and many other types of systems. In a growing state like Colorado, systems-oriented jobs are classified as difficult to hire all over the state because the ratio of demand (available jobs) to supply (of candidates) is low and, in several cases, less than one. Understanding the causes of the shortage and how it behaves over time is a systems thinking exercise by itself.

This chapter lists the many reasons for the shortage of critical systems-oriented skills cited in the academic and professional literature. Using a classical cause-and-effect approach, we visualize these causes on a fishbone diagram. This summarizes the first-order direct or linear chain of causes but does not show their interplay and feedback that drive the shortage behavior over time.

A causal loop diagram is then presented that sheds light on the dynamic behavior of systems thinking skill shortage over time. This highlights the importance of practicing systems thinking and its advantages over the classical way of dealing with cause and effect.

This chapter wraps up with a reflection that explores systems thinking applied to itself (ST^2), focusing on concerns about whether proposed solutions may inadvertently exacerbate issues. By integrating human and learning principles, we compare our perspectives on ST^2 with those of other authors in the field.

8.1 Introduction

We performed a supply versus demand analysis of ten systems-oriented jobs using a 2023 snapshot of Colorado jobs data. This analysis aimed to understand the severity of the shortage of systems-oriented jobs in Colorado to help create sustainable talent

K. Eftekhari Shahroudi et al., *Practical Systems Thinking*,
https://doi.org/10.1007/978-3-031-89394-0_8

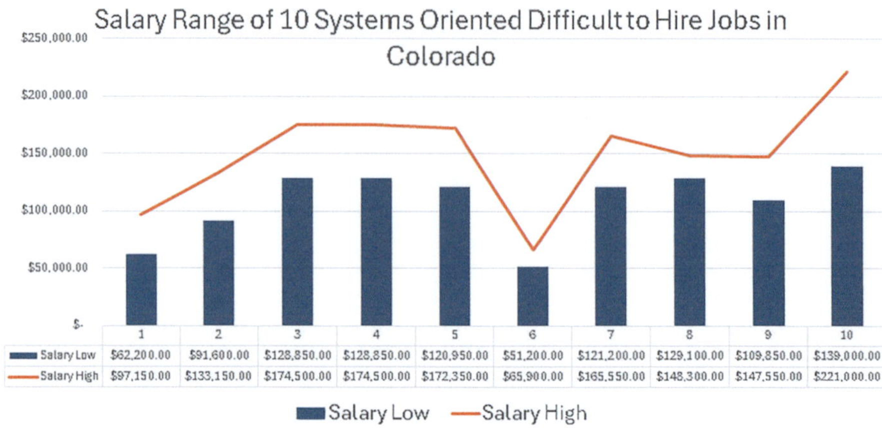

Fig. 8.1 2023 snapshot of the salary range of ten systems-oriented jobs in Colorado

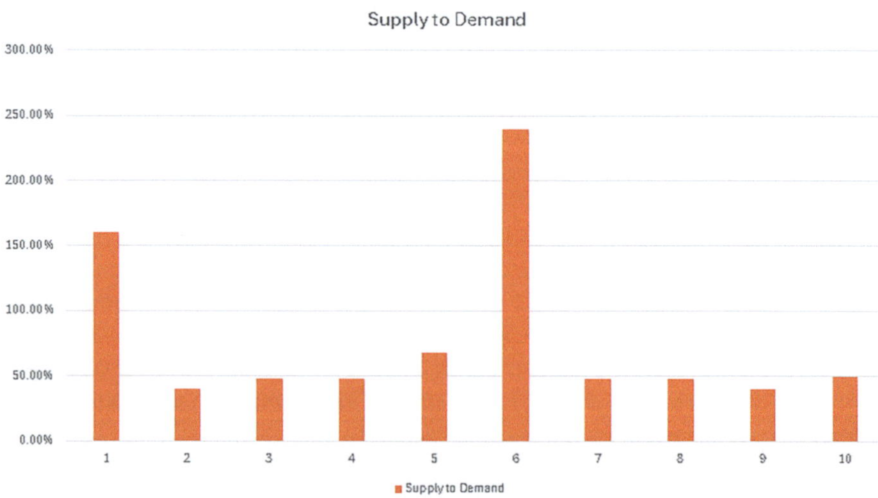

Fig. 8.2 2023 snapshot of the supply to demand ratio of ten systems-oriented jobs in Colorado

flow pipelines. The charts are shown in Figs. 8.1, 8.2, 8.3, 8.4 and 8.5. They suggest the following conclusions for systems-oriented jobs:

- **Great salaries** with several whose lower bound exceeds five figures!
- **Low supply to demand** with several showing less supply than demand
- **Low barrier to entry** with many jobs requiring junior level experience with bachelor degrees
- **Low diversity** where ethnic and female diversity is low for systems-oriented jobs

Not shown in the charts is that most of the available systems-oriented US talent is concentrated on the West and East coasts, with a severe shortage in the

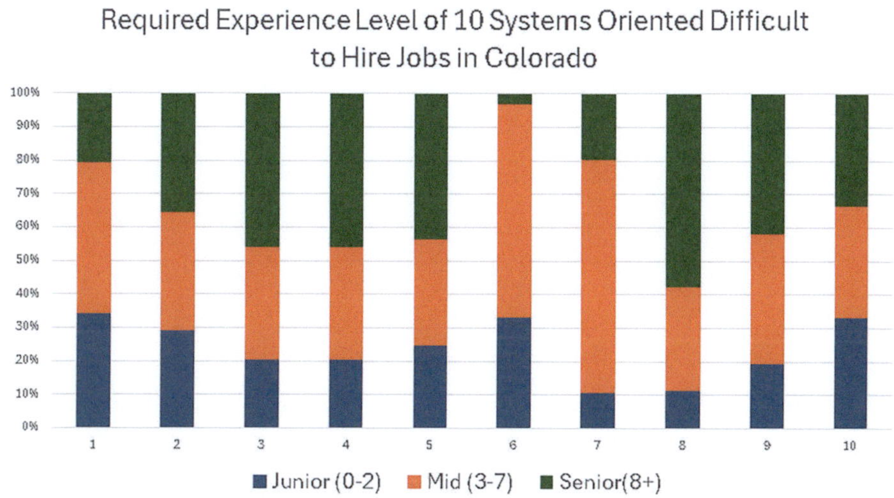

Fig. 8.3 2023 snapshot of the required experience of ten systems-oriented jobs in Colorado

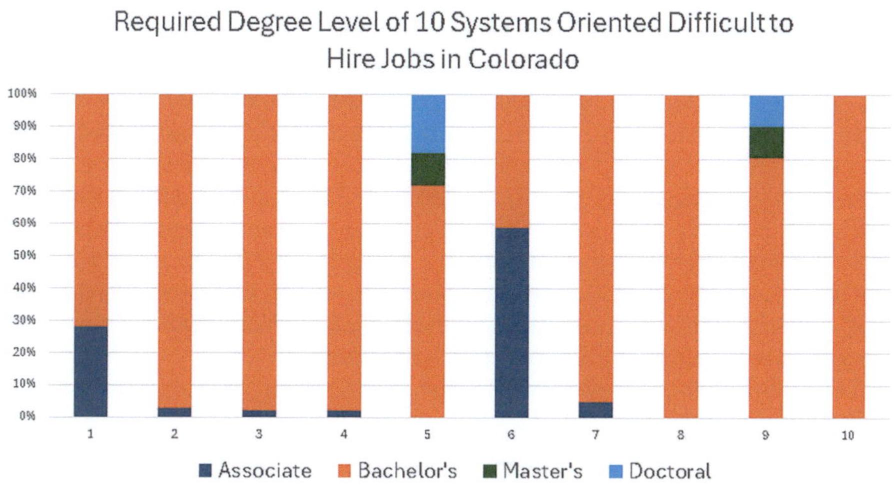

Fig. 8.4 2023 snapshot of the required education of ten systems-oriented jobs in Colorado

middle. So, fresh talent that is already attracted to the central states like Colorado, Arizona, and Texas is likely to find excellent career opportunities there if it develops systems-oriented and systems thinking skills. Our data here is not broad enough to extrapolate our findings to other states.

Nevertheless, Stoewer and Nichols [14] wrote a predictive white paper on the *workforce of the future* listing many systems-oriented jobs in several industries that are predicted to be in high demand till 2035. Our interpretation of their guidance for the next decade is:

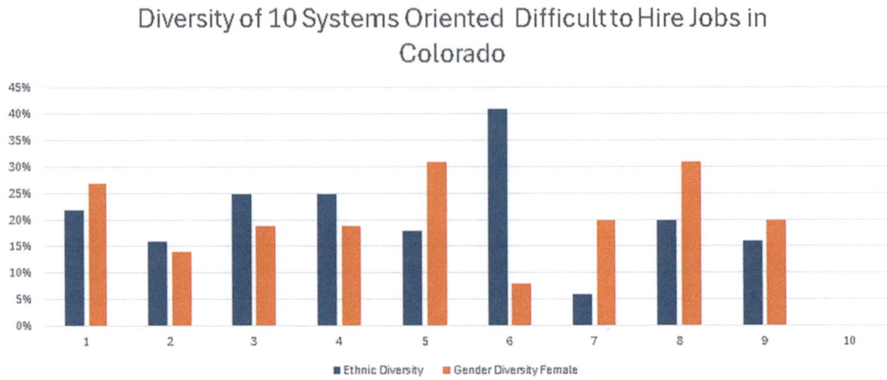

Fig. 8.5 2023 snapshot of the diversity of supply for ten systems-oriented jobs in Colorado

– Most if not all product or service role needs to exercise systems thinking.
– Multidisciplinary knowledge, training, and education are essential.
– Classical disciplines will find a path to become systems oriented.

We acknowledge that Figs. 8.2, 8.3, 8.4 and 8.5 show a snapshot of Colorado industry jobs at one par. The data is not broad enough to include seasonal, regional, and national effects. Also, the reader may wonder why we only chose ten job categories when there may be more than ten that require systems-oriented skills. Our reason was that we were confident that the ten job categories we chose required a significant level of systems-oriented skills, as it is not yet common practice to explicitly list systems thinking and other systems-oriented skills in job announcements. Indeed, we do not believe hiring managers in mature industries like the aerospace industry in Colorado commonly recognize that systems thinking is an essential skill across most roles. Nevertheless, we have been careful not to extrapolate our findings to all job categories and regions.

The following sections discuss the causes for the shortage of systems thinking skill sets generally agreed upon by academic and industry professionals. The first gap contributing to this shortage is the struggle to understand human behavior. The second gap is that our educational system does not teach systems thinking early enough or at all in some cases. The third gap is the methods for teaching systems thinking. The fourth gap is that while multidisciplinary teams are shown to enhance both the learning and application of systems thinking principles, they are not being consistently and intentionally formed. Finally, the fifth gap is notable resistance from global leaders and organizations to adopt systems thinking due to entrenched mindsets.

8.2 The Causes for the Scarcity of Systems Thinking Expertise

8.2.1 Humans Struggle to Understand Human Behavior

People view the world through different lenses. This is true in two ways: there are tangible lenses on our eyes, and then, there are the intangible lenses through which each human sees the world. Tangible means you can touch it, and, in truth, you could touch an eye lens if you dissected an eye, but you could also take your optometrist's word for it that a tangible lens exists on each of your eyes and avoid the mess. Most humans are born with lenses in both eyes, but it is worth noting that there is a rare condition called aphakia, where a baby is born without a lens in one or both eyes. The function of the eye lens is to bend and focus light onto the retina. One of the most amazing things about eye lenses is their ability to automatically change shape to focus on objects at different distances. Thankfully, we as humans do not have to think to cue our eyes every time they need to change shape; it just happens. Most people are also familiar with the common eye problems of nearsightedness, farsightedness, and cataracts. A lens-related cause for nearsightedness is an abnormal shape in the curvature and/or thickness of the eye's lens. A lens-related cause for farsightedness is that the lens is too flat or sitting too far back in the eye. A cataract is a clouding of the lens in your eye that may happen when aging or injury changes the tissue that makes up the eye lens.

So, how do the differences between our tangible eye lenses and those in the intangible lenses we use to view the world? Our default lenses are the lenses we are born with. Approximately 35% of adults in the USA have 20/20 vision without corrective lenses [5]. 20/20 is average eyesight. You can draw your own conclusions from this comparison, but many would argue that systems thinking is not a skill set the average person has. Like those with nearsightedness, some humans naturally have a more reasonable and less imaginative lens they use to view the world. It takes awareness and intention to overcome shortsightedness. Like those with farsightedness, some of us naturally have a dreamer-lens, and it takes effort to also adopt a practical lens. We also have subjective lenses, in addition to our default lenses, that we use to view the world. Just like cataracts, our subjective lenses are influenced by our life experiences. The truth is that things happen in our lives that cloud our ability to understand one another.

Corrective prescription lenses are the most conventional treatment for nearsightedness and farsightedness. Currently, the only way to improve cataracts is to surgically remove and replace the clouded lens with an artificial lens. These methods are effective, but there is a cost for corrective prescription lenses, and there is a cost and procedure required to fix cataracts. The same rings true for changing the intangible lens we use to view the world—it has to be a choice, and we must make an investment for it to work.

8.2.2 There Are Gaping Holes in Our Educational System

The thing about gaping holes is that they are known. Gaping literally means wide open, easy to see. You get the point; these are not pinpricks. We are not presenting unknown information here; it is just a summary of why we have these gaping holes and what is being done to fill them. The first gaping hole is that educators cannot teach systems thinking if they have not been formally trained [7]. Educators at all levels have requirements for continuing professional education (CPE) or continuous professional development (the terms vary by state) hours. CPE significantly improves teachers' professional competence [10] and is already a well-established expectation among the academic community. Choosing formal training in systems thinking and achieving CPE credit for that training is low-hanging fruit.

Another gaping hole is the challenge of finding effective off-the-shelf materials for teaching systems thinking [7]. In industry, if something is not your core competency, then you outsource. Teachers already have a long list of core competencies, and while creating age-appropriate materials for teaching systems thinking is something they are more than capable of doing, asking them to do this pulls them away from their value field. Award-winning author Gwendolyn Galsworth asserted that there are two value fields: "Type 1 is your primary value field, where work happens. Type 2 refers to your secondary fields: functions that support the work but are not the work" [6]. For most educators, creating custom educational materials is a Type 2 value field. Educators see the worth of teaching systems thinking but need ready-made materials. Providing a practical off-the-shelf resource for busy teachers is one of our motivations for this book.

A gap found in traditional engineering education programs is that systems thinking is "perceived to be a substitute or replacement for (as opposed to an adjunct to) conventional, tried-and-true engineering approaches and engineering curricula are full and cannot accommodate additional courses" [11]. A traditional engineering education provides students with a linear perspective, basic equations, design paradigms, standard procedures, and rules of thumb but fails to provide a broad systemic perspective or teach holistic thinking [11]. Seeing teaching systems thinking as a whole new course or as a replacement course is daunting for educators. The first step in successful teaching systems thinking is to integrate it into existing educational frameworks.

8.2.3 We Use Teaching Methods That Do Not Work

Integrating systems thinking into our existing educational frameworks is half the battle. The other half is teaching it effectively. At California Polytechnic State University (Cal Poly) in San Luis Obispo, "learn by doing" is more than a motto, it is a core component of the curriculum. Learn by doing means "learning from experiences resulting directly from one's own actions, as contrasted with learning from watching others perform, reading others' instructions or descriptions, or listening to others' instructions or lectures" [12]. Systems thinking can feel like an abstract concept if there is no action on the part of the students. Educators need to supplement traditional lectures with hands-on exercises and real-world applications.

A 2017 survey found that practical ST engagement is strongly linked to age and work experience, so young students with limited experience need hands-on learning to boost their cognitive development in ST [3]. The Change the World Challenge started in 2021 at Cal Poly with the goal of a providing that hands-on opportunity for learning systems thinking "because when you learn by doing, you're ready day one" [1]. The Change the World Challenge is a transformative, multidisciplinary design competition open to upper-class students of any major. All participants are strategically grouped to ensure cross-discipline representation, with the goal for each team to develop a proposal that addresses some of society's most pressing contemporary problems. Teams are mentored throughout the year by faculty members with diverse academic backgrounds. Final proposals must include a technical element and consideration for broader social impacts. Teams must demonstrate that they have utilized a systems thinking approach in developing their solutions, balancing technical concerns with social, cultural, economic, historical, political, ethical, and environmental considerations. Proposals are presented at the end of the academic year to a panel of industry judges from diverse backgrounds.

The Change the World Challenge is an optional extracurricular activity, meaning it is not integrated into the standard curriculum. However, a scaled-down version of this activity could be utilized in various courses across all colleges Simply asking students, "How would you solve this problem?" in a specific context, followed by teaching systems thinking principles and coaching them on using some of the systems thinking tools, is a great start to integrating systems thinking education into our existing curriculum.

8.2.4 It Is Hard to Foster Collaboration Across Disciplines

Historically, organizational success is measured using the financial metrics of efficiency, growth, and profit [9]. However, the use of a balanced scorecard (BSC) methodology now has traction. The BSC methodology is a management innovation to show how nonfinancial measures connect with financial measures [8]. The nonfinancial measures are then more tangible for the rest of the organization to manage. Overall, the goal of the BSC methodology is to drive a more holistic understanding of the organization as a socioeconomic system. Common metrics on a balanced scorecard add, in addition to financial metrics, a metric for understanding employee satisfaction, customer satisfaction, quality, and productivity. While it sounds like common sense to track these metrics, data shows relatively little use of the BSC methodology due to the need for human and financial resources to make the additional data collection happen [2]. All this being said, standalone financial metrics and the expanded nonfinancial metrics in use on most balance scorecards

Fig. 8.6 Company best practices and classical processes such as the balanced scorecard can be augmented with cross-disciplinary collaboration metrics to improve the systems thinking culture

don't necessarily drive collaborative behavior in an organization. It has been said that "what gets measured gets managed." It will continue to be difficult to foster collaboration across disciplines unless we put in place the appropriate measures. Some suggestions for measures include the following:

- Diversity of Contributions: measure how many people or departments from different disciplines contributed to a project
- Utilization of Shared Resources: measure the use of shared tools and technologies
- Interdisciplinary Problem-Solving: measure the degree to which solutions incorporate insights from multiple disciplines
- Mutual Learning: track the extent to which participants feel they are learning from other disciplines, which can be assessed through surveys or self-assessments

Adding even one of these measures to the balanced scorecard is a step in the right direction toward promoting collaboration across disciplines as depicted in Fig. 8.6. Of course, we also need to get more organizations using a balanced scorecard!.

8.2.5 The Age-Old Problem of Resistance to Change

The Big Book of Six Sigma Training Games [4] has one of the most effective change management games that provides a hands-on experience for understanding resistance to change. The basics of the "Change Your Look" game are as follows: the facilitator tells everyone in the room to stand up and find a partner. Each pair stands back to back. No need to touch one another. A volunteer from each pair identifies as team member A. The facilitator instructs this team member to make five (keep it appropriate) changes to their physical appearance. Team member A may remove a shoe, add a jacket, change the location of their new badge, braid their hair, change their part, flip their glasses upside down, roll their sleeves, etc. When all the A's have had a chance to make their changes, the facilitator then tells the pairs they can turn around and face each other. Team member B must identify all the changes team member A made. Then, it's team member B's turn. The pairs turn back around, and team member B makes five changes to their appearance. Team member A then must identify these changes. At this point, everyone is chuckling and thinks the activity is over. Then, the facilitator throws in a curve ball and tells the pair to stand back to back once again and asks team member A to make five more new changes to their appearance. You'll hear a collective groan at this point, but they'll do it. Finally, after team member B has identified the five new changes, the facilitator asks everyone to take a seat and debrief. The facilitator asks the participants these questions: how did it feel to be asked to make so many changes? What are some things that make people resistant to change? What can you do to make it easier for people in your organization to accept change? The main takeaways from this discussion are usually as follows:

- We did not know why we were making these changes.
- The facilitator did not give us a heads-up that we were going to be asked to make changes.
- The facilitator did not give us any training or support to easily make these changes, i.e., the facilitator did not provide examples on how participants could change their appearance, the facilitator did not prepare them ahead of time to over-accessorize for the day, and the facilitator did not give them the materials (sunglasses, hair ties, wigs, hats, etc.) to easily make the changes.

From this exercise, participants learn that if they want to facilitate sustainable change, they must first start by explaining why the change is needed, intentionally be the change they want to see in the organization (utilizing the principles and tools in their daily practice), provide an abundance of communication that demonstrates how systems thinking was used to determine the coming change, and develop a strategy for effectively supporting the integration and sustained change. The same approach applies to overcoming resistance from global leaders and organizations in adopting systems thinking: be like Simon Sinek and always start with why [13], integrate the use of systems thinking principles and tools into your daily practice, and create an environment that makes it easy for leaders to adopt a systems thinking culture.

8.3 Comparison of Tools to Analyze Causes

8.3.1 The Cause-and-Effect (Fishbone Diagram) and Classical Application

The linear cause-and-effect diagram, also known as a fishbone diagram due to its resemblance to a fish skeleton, is a tool used to identify and organize potential causes of a problem. Here's how it works: the problem is placed at the "head" of the diagram, and each major "bone" is labeled with categories such as Human, Machine, Method, Materials, and Environment (sometimes referred to as "Mother Nature"). For each category, brainstorm potential causes, gathering input from stakeholders to ensure diverse perspectives. The purpose of the exercise is to expand your view beyond your current paradigm. While you may already know some causes, a comprehensive analysis ensures that all possible factors are considered. A cause-and-effect diagram for the shortage of systems thinking expertise would look like Fig. 8.7. Once the cause-and-effect diagram is "complete," the next step in any process or quality improvement project is to analyze the diagram. Begin by eliminating causes that are beyond your control and then highlight those that represent root causes, which are provable, concrete facts. If some causes are legitimate but not the most fundamental, conduct a "5 whys" analysis, asking why repeatedly to uncover the lowest level of fact, and gather data to validate it.

There are several variants of this problem-solving approach. This process of finding causal chains has been used repeatedly for various quality improvement projects. However, it has its roots in linear compartmentalized thinking. It is, for example, not able to shed light on emergent behavior that results from the interplay and interaction of different causes and effects with each other, such as reinforcing or balancing loops that drive emergent behavior over time. The next section uses a causal loop diagram to shed more light on how the shortage of systems thinking skills behaves over time.

8.3.2 The Causal Loop Diagram and Systems Thinking Applied to Itself (ST^2)

To better understand the complex behavior of the availability of systems-oriented talent, we started to search for and identify:

- **Relations between causes on the fishbone**: as opposed to the earlier assumption that all the causes on the fishbone diagram of Fig. 8.7 are independent
- **Relations between the effect (skills shortage) feeding back to the causes**: as opposed to assuming that all the causes end with the skills shortage

After several iterations and discussions, we visualized our findings using a causal loop diagram that captures how the systems-oriented skills shortage feeds back to the causes we had identified on the fishbone diagram earlier. The blue part of the causal loop diagram in Fig. 8.8 shows the result: the "availability of systems-oriented talent" from academia is naturally governed by at least four reinforcing loops—**collaboration, training material, teaching methods, and change management** reinforcing feedback loops. As described earlier, reinforcing loops amplify or boost changes in the same direction. So if the "availability of systems-oriented talent" is moving in a downward direction, then these loops all work together to accelerate its fall like gravity!

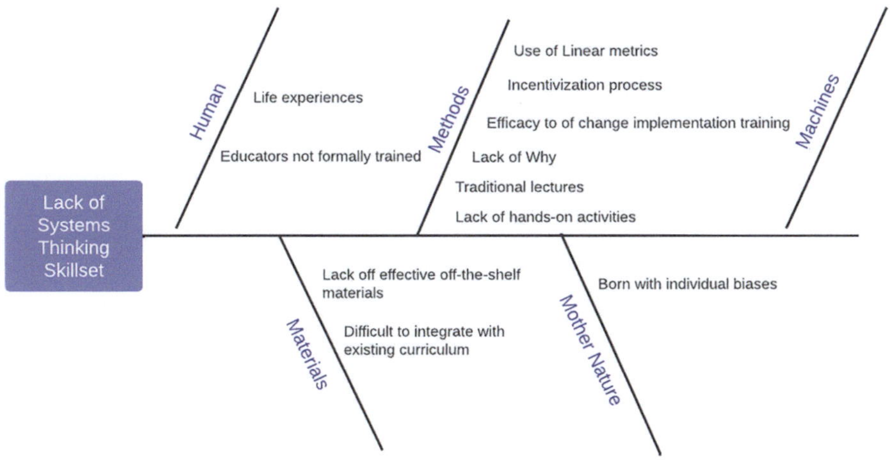

Fig. 8.7 Traditional cause-and-effect (fishbone diagram) for shortage of critical systems-oriented skills

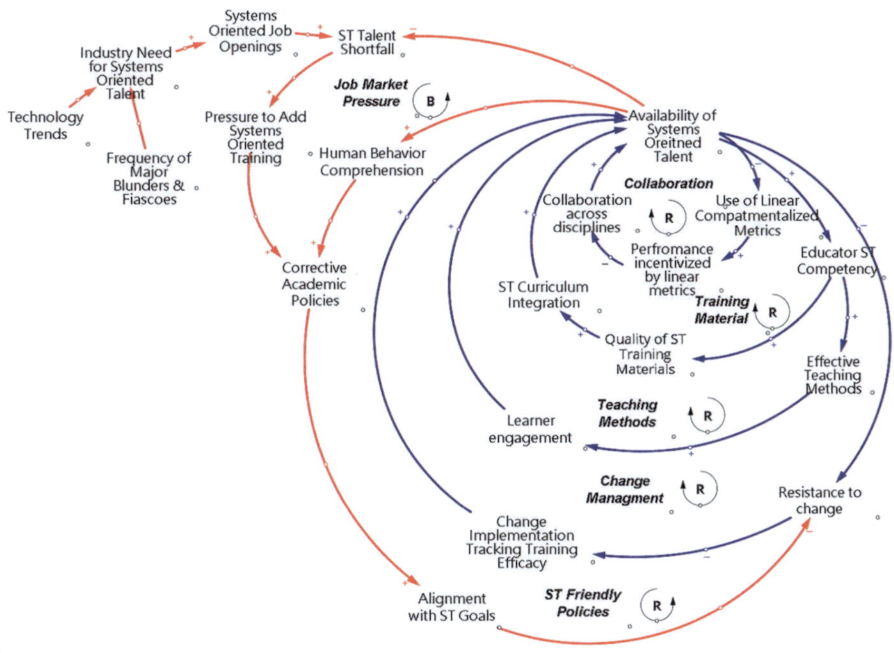

Fig. 8.8 Understanding the complexity of *systems-oriented skills shortage* using a causal loop diagram

At this point, we should start seeing some advantages of systems thinking using causal loop diagrams over the classical approach to identify independent linear chains of cause and effect. Systems thinking encourages us to approach the real-world problem of the "availability of systems-oriented talent" using the *systems principles* that help us better understand its complexity. In this case, the *humans principle* encouraged us to identify all the individual causal relationships between all the key human-related variables that may influence the complex behavior because it states:

Human Principle: *Humans* impact and are impacted by the (wanted/unwanted) *emergence* of *all* reachable systems.

Furthermore, the *emergence principle* encouraged us to identify circular interactions or feedback loops that may influence the complex behavior because it states:

Emergence Principle: *Interactions*(including feedback) within and between the 3-systems cause (sometimes unwanted) *emergent behaviors*.

Note that even the **change management** loop in Fig. 8.8 is reinforcing. So, if the "availability of systems-oriented talent" increases, this loop reinforces the availability to accelerate even faster. However, if the "availability of systems-oriented talent" is dropping, then this loop causes the availability to drop even faster. This was a counterintuitive result that change management can actually make the problem worse!

A similar counterintuitive result was adding the red **ST friendly policies** loop which captures the behavior of the academic leadership trying to put incentives and policies in place to align the organizational culture with systems thinking. Again, this is a reinforcing loop, and despite their best intentions, academic leadership attempts to improve the culture can backfire and make the situation worse if the "availability of systems-oriented talent" is dropping. The reason for this is that the **ST friendly policies** loop works through "resistance to change" and not directly on the "availability of systems-oriented talent".

So is the "availability of systems-oriented talent" doomed once it starts to drop? Here, the *3-systems principle* guided us to look outside the *system of interest* that was the shortage of availability of systems-oriented talent from academia because it states:

3-Systems Principle: *Systems* come in (at least) *3s—context systems*, *systems of interest (SoI)*, and *enabling systems*.

When we looked outside the *system of interest* to understand if there are interactions between the *system of interest* and the *context system* that influence the complex behavior, we discovered the impact of "industry need for systems-oriented talent" that continues to create "systems-oriented job openings". The "ST talent shortfall" then creates pressure on "corrective academic policies". Figure 8.8 shows that this sets up a balancing loop called **job market pressure** that forms a goal-seeking behavior to ensure that the "availability of systems-oriented talent" is always pushed to meet the required "systems-oriented job openings": in other words, regardless of whether "availability of systems-oriented talent" is increasing or decreasing, the **job market pressure** loop always tries to exert pressure on academia in the correct direction to meet the industrial demand. So, the "availability of systems-oriented talent" is not doomed once it starts to drop. Now that we have found this first goal-seeking behavior, we can find additional similar loops that come from influencers or evangelists from industry consortia such as INCOSE, PMI, AIAA, ASME, IEEE, SAE, etc.

We are not claiming that we have fully understood the complexity of the problem of the availability of systems-oriented talent. Yet, having identified the reinforcing and balancing loops that influence it makes the problem relatively easier to grasp and more predictable than before. None of the structural insights discussed in this section would have been possible with the classical approach to identifying linear chains of cause and effect or creating fishbone diagrams. We are not against classical disciplines, approaches, or methods that have previously enjoyed some level of success. However, we need to understand their limitations and how to apply, augment, or integrate them into our thinking better. For example, there is nothing wrong with creating a fishbone diagram as a starting point. But then, we should search for relevant reinforcing and balancing feedback effects that a fishbone diagram cannot capture.

Now, assume that our CLD understanding of the lack of systems-oriented talent is correct today, and this understanding has become common knowledge. If we act on it, can we predict that the shortage will disappear in a few years? It depends on how our actions interact with the other systems: context systems and enabling systems. For example, we can experience policy resistance: we might spend resources on improving the shortage in the near term, only to find that the rest of the system that does not align with systems thinking culture may resist our efforts so that over time, we may end up with the same shortage problem, despite our efforts and expenditure to try to move the needle in the right direction. Systems thinking encourages us to find the best leverage point to direct our efforts and anticipate the unwanted consequences over time. But it is a relative advantage and not a guarantee. So, systems thinking can shed light on complex real-world issues, even if the issue is related to itself: the availability of systems-oriented talent!

8.4 Reflection on Chapter 8 Material

In this chapter, we showed evidence of the shortage of critical systems-oriented skills that are in high demand for ten categories of well-paying jobs in Colorado that do not currently have a high barrier to entry. We cited a reference that predicted global demand for systems-oriented skills in many industries until 2035. We collected the different causes of the shortage from cited academic and industry professionals. We then constructed a traditional cause-and-effect (fishbone) diagram shown in Fig. 8.7 to visualize the findings.

The fishbone was a good start but unable to capture the complexities and subtleties related to how we got into the shortage of critical systems-oriented skills and where to direct our efforts to improve it. So, we applied systems thinking to itself (ST^2): we developed a causal loop diagram (CLD), a rigorous systems thinking method, to better understand the reinforcing and balancing loops that impact the systems-oriented skills shortage. Comparing the CLD in Fig. 8.8 to the fishbone in Fig. 8.7 shows that the former enables a richer understanding of the issue and better means of identifying its leverage points. For example, we can use the CLD to see some potential unwanted consequences of a policy that intends to alleviate the shortage. Now, it was not luck that helped us develop a richer understanding of the shortage. We explained this was a consequence of applying two of the systems principles of Chapter 5: the *human principle* and the *emergence principle*.

Several variables in the CLD of Fig. 8.8 are strongly influenced by our personal and organizational systems thinking capabilities and culture. We discuss these in Chapter 9 next. Part III will next introduce us to a range of rigorous methods that will augment our capabilities to apply this philosophy to real-world problems, such as the model-based systems thinking (MBST) method.

References

1. Cal Poly (2024). *Learn by Doing*. Web Page. https://www.calpoly.edu/learn-by-doing
2. Benková, E., et al. (2020). *Factors Affecting the Use of Balanced Scorecard in Measuring Company Performance*. Sustainability
3. Camelia, F., & Ferris, T. L. J. (2016). Undergraduate students'. *IEEE Transactions on Systems, Man, and Cybernetics: Systems, 47*(12), 3165–3176. ISSN: 2168-2216.
4. Chen, C., & Roth, H. M. (2004). *The big book of six sigma training games: Proven ways to teach basic DMAIC principles and quality improvement tools*. McGraw Hill Professional. ISBN: 007-17-88-02-6
5. Cleveland Clinic. "20/20 Vision." 8 October 2024. *Cleveland Clinic*. https://my.clevelandclinic.org/health/articles/2020-vision.
6. Galsworth, G. (2016). *Drilling Deeper Into Your Value Field*. Electronic Article. https://www.linkedin.com/pulse/drilling-deeper-your-value-field-gwendolyn-galsworth-phd-lion-
7. Gilbert, L. A., Gross, D. S., & Kreutz, K. J. (2019). Developing under-graduate students' systems thinking skills with an InTeGrate module. *Journal of Geoscience Education, 67*(1), 34–49. ISSN: 1089-9995.
8. Nørreklit, H. (2003). The balanced scorecard: What is the score? A rhetorical analysis of the balanced scorecard. *Accounting, Organizations, and Society, 28*(6), 591–619.
9. Maltz, A. C., Shenhar, A. J., & Richard, R. R. (2003). Beyond the balanced scorecard: Refining the search for organizational success measures. *Long Range Planning, 36*(2), 187–204.
10. Mallarangan, A. D. D., et al. (2024). Analysis of the influence of continuous training development and education on professional competence of teachers in public schools. *Journal on Education, 6*(2), 13449–13456. ISSN: 2654-5497
11. Monat, J., Gannon, T., & Amissah, M. (2022). The case for systems thinking in undergraduate engineering education. *International Journal of Engineering Pedagogy (iJEP), 12*(3), 50–88.
12. Reese, H. W. (2011). The learning-by-doing principle. *Behavioral Development Bulletin, 17*(1), 1. ISSN: 1942-0722.
13. Sinek, S. (2009). *Start with why: How great leaders inspire everyone to take action*. Penguin. ISBN: 110-1-149-03-5.
14. Stoewer, H, & Nichols, D. (2023). Building the systems engineering workforce of the future: A complement to the incose systems engineering vision 2035. In *Guidance for Education, Training and Development of Systems Engineers*.

Chapter 9
A Culture That Fosters Systems Thinking

Abstract At the time of writing this chapter, there are already many articles, books, YouTube videos, and classes that encourage or teach systems thinking. However, Chapter 8 showed that the supply of systems-oriented fresh talent is low relative to the demand that in some states like Colorado can pay five figures starting salaries without a super high educational or experience barrier to entry. Our observations are that use of systems thinking terminology is growing but linear compartmentalized mental models are still dominant in individuals and organizations. For systems thinking to balance this linear compartmentalized thinking, we need individuals and organizations to adopt a systems thinking culture. Chapter 5 listed the *systems principles* that underpin the systems thinking philosophy. Chapter 6 helped us develop a sensitivity for this philosophy to see the principles in action in our daily lives. Chapter 7 gave the process and behavioral guidelines for applying *systems principles* to real-world problems. This chapter discusses the individual and organizational journey toward a culture that fosters systems thinking ending with a section that discusses how we would potentially measure progress on this journey.

9.1 Personal Journey Toward a Culture That Fosters Systems Thinking

You may have already applied systems thinking unconsciously when solving jigsaw puzzles. To construct a puzzle, we consider the entire puzzle and develop strategies to examine how pieces fit together, how colors change, and how we observe similar patterns. But imagine yourself solving a jigsaw puzzle by only being able to view one piece or one part of the puzzle at a time. How would you know what the final puzzle should look like, and how would you sort similar and dissimilar pieces? You might guess at one piece, and you may find a piece that fits only to realize that it's the wrong piece and you have to start over. You would have a nearly impossible task where solving the puzzle is just a matter of luck.

Unfortunately, this is how many of our most complex problems—climate change, homelessness, and wildfires—are approached. We look at one piece of the system trying to find solutions that fit, but when we step back and view the entire system,

K. Eftekhari Shahroudi et al., *Practical Systems Thinking*,
https://doi.org/10.1007/978-3-031-89394-0_9

we realize that we may have only solved one piece of the puzzle and have yet to address the problem as a whole.

Now, how do we get to a capability where systems thinking becomes almost second nature on a personal level? A work, school, or club environment that fosters or requires systems thinking skills will certainly push us in the right direction, but many of us are working or living in environments dominated by linear compartmentalized thinking and culture. So, here are some core competencies that will need to find along our personal journey toward systems thinking:

- **Personal Mastery of the *Systems Principles***. This requires learning and applying the *systems principles* listed in Chapter 5 to improve our understanding of real-world problems and the systems intended to influence and interact with them.
- **Alignment of Behavior with *Ways of the Systems Thinker***. These are behavioral guidelines (do's and dont's) listed in Chapter 7 requiring us to balance our effort and focus among all the concerns and features emphasized by the Chapter 5. These guidelines are summarized in Fig. 7.2 of Chapter 7.
- **Application of *Rigorous Tools for Systems Thinking***. This is where we are not just talking but also walking! Several easier-to-get started-with rigorous tools such as system mapping and causal loop diagrams were already described in Chapter 7. Then, Part III of this book is dedicated to several deeper rigorous tools that include various forms of modeling such as architectural, causal, relational, AI, and agent-based modeling that enable systems thinking practitioners to apply the *systems principles* to real-world problems or systems that may influence them.
- **Gain Systems Thinking *Experience***. This is learning by doing where we look for any real-world opportunity in our work, study, or home to apply the principles, behaviors, and tools. The following situations might, for example, be a candidate to learn by doing: (1) A problem or an unwanted pattern of behavior that is consistently recurring (such as projects always over budget and late); (2) A problem that has counterintuitive behavior (such as weight gain as a result of dieting); (3) A problem that resists or works against a policy or investment to fix it (such as homelessness and wildfires that continue to get worse despite increased resources directed to improve them); (4) Silos where people conveniently have a very narrow view of the scope of their roles and responsibilities; (5) Picking quick solutions without considering alternatives or their potential unwanted longer-term side effects; (6) Situations where previous solutions become the current problems; and many more real-life situations where previous approaches have repeatedly failed.

We are not suggesting a sequential approach to obtaining the above competencies. So, for example, we are not saying that the best approach is to master all the in-depth rigorous tools and then find a situation to use them. We are suggesting balanced growth along these competencies needed to solve our challenges. The rest of the material in this section is intended to encourage awareness of the way we think and approach problems, highlighting different aspects of our growth on the personal journey toward a culture of systems thinking.

9.1.1 Noticing Systems Thinking in Our Routine

Systems thinking introduces a different way of thinking where we consider the system as a whole and the purpose of the system before looking for solutions. The question is how do we create a culture of systems thinking where all problems are addressed with this holistic mindset? This section examines the concepts of systems thinking and how these concepts foster a culture of systems thinking that can help individuals navigate and address the complexities of our interconnected world.

9.1.2 Systems Thinking Can Evolve from the Bottom-Up

An early application of systems thinking can be found in the nineteenth-century fur trade in British Columbia. During this period, the Hudson's Bay Company controlled and managed a vast portion of the British western territory. Over a period of 90 years, the company bought Lynx and Hare pelts from trappers, and they meticulously recorded where and how many pelts were bought across the territory. These records revealed an interconnected pattern of growth between lynx and hare populations. As more lynx were trapped, the populations of hares grew substantially, causing trappers to shift to hare trapping. Then, as hare populations grew, more food was available causing a growth in lynx populations, and more lynx pelts were bought. This oscillation between lynx and hare trapping continued as long as neither lynx nor hare was overhunted. Indigenous trappers would follow the population shifts; however, Hudson Bay Company records also noted an impact of European trappers that tended to overhunt populations. As trappers caught more lynx, hare populations would grow beyond the capacity of the environment to support, and hare populations eventually collapsed. Likewise, if trappers focused only on hare populations, lynx populations would also decline leading to a reduction in both lynx and hare pelts being purchased. The Hudson Bay Corporation observed these relationships and encouraged Europeans to learn from indigenous trappers to create a more sustainable supply of pelts.

Understanding overall system behavior and its limits is a key principle of systems thinking, and the data provided by tracking animal pelts provided insights into how systems thinking applied to the western fur trade. First, we see the interconnectedness between human and ecological systems in trapping influenced both lynx and hare populations and that these populations influenced the activities and welfare of trappers. Understanding and mapping these interactions is a core component of systems thinking in seeing the big picture. Consider how your personal decisions can impact your family, work, and community and how their actions influence your life. The size and quality of your social network influence the overall interconnectedness of your life. Second, we observe feedback loops at work in the fur trade in how overhunting of lynx reduces the balance (negative) effect on hare populations driving their positive population growth until the available food supply creates another balancing or negative effect to limit the number of hare. Individually, you experience positive and negative feedback through your positive and negative growth experiences. The more positive interactions restaurants have with diners, the more people will choose to eat there; however, as a restaurant becomes crowded and service declines, it will create a negative or balancing effect, and you may no longer or choose to eat there less often. Third, we find boundaries in the fur trade. Local hare populations are affected by local and migrating lynx populations and traders that roam. Boundaries define what is in and outside a system, and we consider boundaries in managing our work-life balance or we might observe how our boundaries change as we add or remove elements from our life. Setting boundaries helps us to think about a system as well as think about how we might need to change the boundary to fully understand its behavior.

If these early recordings of system interactions and behavior help us to see the big picture, then rigorous tools like system dynamics help us to realize the final picture of how all the puzzle pieces fit together. System dynamics models the interactions, feedback, and behavior of the system to predict how the system will behave over time. These models help us to see how we may change the system and how these changes impact the system overall.

9.1.3 From Mental Models to Character

Our thoughts form our behaviours, which form our habits, which eventually define our character. Fostering a good culture of systems thinking begins with changing our mental models. We develop mental models to understand and interpret the world around us. They help us to simplify reality and give us an internal representation of that reality. They help us make decisions, process information, and predict outcomes. Mental models are shaped by our personal experiences, our social interactions, our education, our culture, and how we perceive and respond to different situations. Our mental models are not static, however. They change as we change. The journey toward becoming a systems thinker means changing our mental models of our world. This will involve questioning and updating our beliefs and assumptions. It does not mean that we necessarily need to abandon our existing models but systems thinking takes a holistic view, so individually we will want to foster a culture of critical thinking and connectedness to add to our mental models. Shifting our underlying mindsets can transform the entire system. Challenge yourself and your organizations to add additional mindsets. History is flush with inventions and discoveries achieved by changing our mindset of the problem.

Once such story is in the discovery of antibiotics. After a long holiday in late 1928, Alexander Fleming began to sort through his petri dish colonies containing Staphylococcus. Staphylococcus is a bacterium that leads to sore throats, boils, blisters, and redness on your skin. He noticed that one dish was contaminated with a mold patch and there was no bacteria growth around the mold patch. Fleming experimented and found that mixtures with this mold—identified later as Penicillium notatum—was capable of killing several forms of bacteria.

Before this discovery, bacterial infections were viewed as incurable. Had Fleming held to his and societies mental model and discarded the petri dish with mold as contamination, we would not have discovered antibiotics. So, his shift in thinking has revolutionized medicine and saved millions of lives.

Creating a culture of systems thinking means fostering continual learning and working toward sharing a common vision of the world informed by a systems view. Systems thinking is about coming together so strive to foster a collective understanding and learning process. Encourage and learn from your peers and people with different viewpoints. Organizations can support collaborative learning environments where group problem-solving is standard. This process starts by aligning our personal goals with a larger purpose, and organizations can create a holistic and compelling vision of the purpose of the company. As a system is a collection of entities that come together for a specific purpose, sharing and committing to that purpose will help the system perform as intended. Moreover, changing the purpose and goals in the system can dramatically impact outcomes.

You can start by applying the tools in this book to understand how the systems you part are of behave and then look for aspects that exhibit goal-seeking behavior. Set new goals for those behaviors. Start small, setting personal goals in realizable aspects of your life, for example—how much you spend on eating out in a month or how you organize your email. Look for how other aspects of your life change as you approach your goals. Be careful in setting goals as not achieving goals creates a downward spiral called collapsing goals where we continually set smaller and smaller goals to achieve them.

9.1.4 Personal Mastery and Power to Transcend Paradigms

By adopting the behaviors of systems thinking, you give yourself the power to transcend paradigms by starting to see the world around you in a completely new perspective. You can practice mindfulness and challenge yourself to look for additional connections in the systems around you. First, start by recognizing that you are in and are part of a system. You have a social system in your family and friends. These social systems intersect with your work systems and colleagues. You are then part of a community system. Practice looking for three connections to you and the elements of your life. Do the same at work and encourage your colleagues to adopt a strategic mindset. You will not be a systems thinker overnight, but once you begin to see the interconnectedness of life, you will never be able to follow siloed thinking again. Don't be frustrated and allow yourself time for continual learning and self-improvement. Strive for personal growth. Doing so will benefit you, help you to be recognized, and succeed at school, work, and in social settings. So, let's look at how we can start to create a culture of systems thinking through a three-system mindset.

Leverage Points
Every system is driven by the components and interactions that come together for a specific purpose. Changing the quantities and size of the components can influence the behavior of the system. If we increase the number of cashiers at a grocery store, we can reduce the time people wait in line. Or if we decrease the number of fish that can be caught, we reduce the impact of humans on fisheries around the world. But often, these changes are not enough to solve problems. We can exert greater leverage on the system by understanding the root causes and system structures that drive behavior. Just increasing the number of electric vehicles on the road, for example, does not change the underlying system that is reliant upon fossil fuels, but it will ease pressure on that system while we look to find more substantial solutions. We can apply this thinking to our daily routines by making small differences while pursuing more substantive changes.

Buffers and Delays
Our bodies have many buffers that create a state of homeostasis, the steady state of function maintained by living organisms. We store fat and process energy slowly so we can manage the periods between eating, a good thing for being able to work and enjoy life for sure. Many systems have buffers that delay or dampen the effect of change. Buffers act like a reservoir that help keep systems functioning. Buffers also disguise the benefits from some solutions. Dieting is challenging as it takes time to realize the benefits. A decision today to invest in education may also take years to change the system. Modifying the size of buffers can reduce the time it takes to observe any effects; however, it affects the stability of the system. Losing too much weight can affect the homeostasis of our bodies. Considering that all systems have some buffers helps us to build up reserves to manage uncertainties as well as appreciate the time required to see change.

Stocks and Flows
The world's dynamic systems are organized by stocks and flows. Stocks are those items in the world that build quantities like water reservoirs, bank accounts, and populations. Flows are those structures that move stocks around, like rivers, spending, and births. When we change the rate of flow or the structure of the system, we change the flows between stocks. Health departments around the world asked us to limit our contact with others to keep infections low during the COVID-19 pandemic. Reducing contacts reduces the flow of people from stocks of "healthy people" to "infected people." If we change the way we interact with people (going from in person meetings to online meetings), we change the flows altogether, and we can create new systems. You can consider these stocks and flows in your daily schedule by redesigning your day, your workflow, to achieve your goals.

Rules and Self-Organizing Behavior
Spend some time observing a public square or train station and you'll start to notice how people move around each other. There are few signs that tell us to not run into people, walk on the right or left, hold the door for people, and let bikes go by, and yet, there are societal norms or simple rules of life that guide the systems around us. These rules influence how a system operates, and you can establish personal rules or guidelines to stay focused on the purpose of the system and or achieve goals. Rules also lead to self-organizing behavior. Consider a community playground. Bring a group of children to the playground, and within a short time, you will see patterns and group play emerge. It will not matter whether the children know each other beforehand as there are some simple rules that guide how the play will organize. We see self-organizing behavior in all systems, and we can alter the rules, laws, and incentives to change how the system functions. Online multiplayer game developers consider rules and self-organizing behavior when developing popular games such as Fortnite, Call of Duty, and Roblox. Successful game developers have also recognized that rules must not be rigid as giving a system the ability to evolve its structure can lead to more resilient systems. Resilient systems can foster innovation, adaptation, and a mindset to adapt to life's uncertainties.

Feedback Loops
Earlier, we looked at the effect of reinforcing and balancing feedbacks in the lynx and hare pelt trade. Strengthening your awareness of feedback loops will help you consider where and how things in life and your organization go askew. Watching for and encouraging balancing loops will help keep systems stable. Taking time to relax at work or school can help maintain well-being so that reinforcing stress does not grow out of control. Organizations can implement monitoring and health awareness programs to ensure product quality and consistency. Or if you aren't experiencing enough personal growth, you can introduce reinforcing loops or remove balancing (negative) influences that are limiting growth. Reinforcing feedback loops do not only drive systems in an upward direction. Look for reinforcing loops that drive paths downward as well. Reinforcing loops are self-perpetuating so if not controlled they will reinforce a particular behavior until the system collapses. Look for habits, relationships, and performance that are spiraling downward to intervene and add balancing relationships to slow down or control this behavior.

Interconnections and Information Flows
Over 7 billion people who have smartphones bring a wealth of information to us all. The interconnections between system elements are the information highways. The control and flow of information substantially impact a system's behavior. Many companies experienced a dramatic shift in their supply chain during COVID-19. These shifts were caused by a failure of information between consumer and suppliers of goods and materials. Consumers started hoarding goods which created a fluctuation in consumer demand at the retail level. Suppliers responded by purchasing more goods, which in turn caused progressively larger demands at the wholesale, distribution, and manufacturing levels. These fluctuations, termed the bullwhip effect for its similarity to the way a whip moves at the end given very

small movements at the handle, illustrates how a small change in information can cause large effects. And while consumers did purchase more goods, these purchases represented short-term behavior and not a sustained increase in demand. Companies responded by overproducing goods, utilizing more resources which in turn caused supply issues in other manufacturing lines. By providing more information to the people and systems you are working with, you can stabilize system behavior. Individually, this means improving communication in your personal network, and for organizations, it means improving the transparency and flow of information so that internal decision-making can respond to accurate information.

Viewing Systems in Nonlinearities
Many parts of our complex systems behave in nonlinear ways. Population growth is exponential in that it is a reinforcing loop that drives more and more births. Compound interest is also nonlinear in that the more savings you have the more interest is gained. Early systems thinkers recognized these nonlinearities in the system: first in recognizing the reinforcing loops that could spiral out of control and, second, in recognizing that nonlinear behavior can lead to complex and unpredictable behavior. When 2+2 equals 2+2 * some growth rate, predicting the impact of decisions becomes more difficult. Recognizing the parts of a system that behave this way is a way to identify and reorganize the actions that may not have straightforward outcomes. Organizations can plan accordingly and prepare for nonlinear responses.

9.1.5 How Rigorous Tools for Systems Thinking Encourage Our Learning and Growth

Section 12.2.3 gives an example application of system dynamics that helps us to consider a holistic view of our world. System dynamics has evolved over the decades and system dynamics have been applied in business to improve manufacturing, applied in policy analysis to study the impact of transportation policies in Oregon, and applied to model world climate to address environmental challenges and sustainable development.

There are free tools like Vensim PLE (a free downloadable platform for personal use) and Insightmakerċom (an online platform with thousands of examples) to start learning system dynamics and visualizing how systems interact and change. System dynamics is also taught in many universities and colleges where you can gain an introduction to the concepts and tools of systems thinking or study advanced applications. You can look for courses such as:

1. **Introduction to System Dynamics or Systems Thinking**: Courses that introduce system dynamics or systems thinking provide a foundation in understanding how systems work. They are designed to introduce basic concepts, tools, and techniques for using system dynamics to model complex systems.

2. **Advanced System Dynamics**: Advanced courses provide a more in-depth exploration of modeling techniques and applications in various fields. Advanced courses teach students how to apply system dynamics to specific areas of interest. Advanced courses may also focus on problem-solving or decision-making.
3. **System Dynamics in Specific Context** (e.g., business, policy, environmental modeling): Look for courses in a specific context to gain learning on how you would use system dynamics to evaluate and develop public policies and model the impact of policies, leverages, and changes on system behavior. System dynamics for business courses will focus on how to address business challenges and improve organizational performance. Business-focused courses help students with careers in business analysis, operations management, and strategic planning. Environmental modeling courses will focus on applying system dynamics to water, energy, and ecological systems.

Learning and applying system dynamics will help businesses optimize all aspects of their internal systems from supply chains to project management. It will also help employees understand market dynamics. Learning these tools will help you differentiate yourself, and organizations can benefit from training their employees. With a culture of systems thinking and the system dynamics toolset, organizations can leverage these skills to optimize business processes and strategic initiatives. In Part III of this book, we detail several tools that support a culture of systems thinking and core tools and techniques, including system maps, causal loop diagrams, mind mapping, stock and flow diagrams, and simulation models.

You can begin to foster a mindset for systems thinking by internalizing the key principles of systems thinking and system dynamics on a personal level. Start by mind mapping small actions in your life. First, start with a central idea or problem you are facing and branch out to related concepts, influences, actors, or systems. Draw your central idea in a circle and use lines and arrows to show connections. Brainstorm until you have reached at least three system levels (three degrees from the central idea). Second, highlight the most important elements and relationships asking which connections are needed to support the central theme. Look for a grouping of ideas and connections that may suggest potential feedback loops. Third, consider how different elements interact and influence each other. Add active words to each connect indicating things like the link is supportive, decreases, adds to, and reduces between elements. Finally, identify any potential points of intervention or leverage.

From mind mapping, you can start to create formal causal loop diagrams. Recall that casual loop diagrams also map system elements and interactions but focus on documenting the behavior of the system and the feedback loops that drive the system. Preparing a casual loop diagram can illustrate how different elements influence each other and where the system has balancing and reinforcing behavior that needs attention. As with mind maps, preparing a causal loop diagram starts with a key variable or element in the system. Focus on the stocks in the system, the people, the technologies, and the processes. Then, draw arrows to show the direction of influence, indicating what elements cause an effect in another element.

For instance, to understand how population pressures affect housing shortages, you might draw two stocks, people and available housing. Then, draw an arrow from people to available housing. As more people reduce the number of available houses, you would label that arrow as decreases indicating that as population increases, available houses decrease. With casual loop diagrams, we label this as a (-) to indicate a reversal of direction. Continue working through the system of interest until you have documented all elements and interactions.

Next, look for feedback loops in the diagram. These might be reinforcing or balancing. Focus on how these loops affect the behavior of the system. Once you have a draft casual loop diagram, you can then explore different scenarios and predict how the system would change as you changed one or more elements in the system.

9.2 Organizational Journey Toward a Culture That Fosters Systems Thinking

According to Google Search labs: *Organizational culture (OC) is the set of shared beliefs, values, and expectations that guide the interactions and behaviors of people within an organization. It's often described as the personality of an organization, and it can affect how employees, customers, and others experience the organization.* There are many other definitions, but there is a strong agreement in the literature that organizational culture is influential as it strongly influences the design and application of tools, processes, and organizational machinery to solve their multifaceted challenges. So, the chain of command, how major decisions are made, acceptable codes of behavior, beliefs, values, dominant language, and terminology in an organization are all strongly influenced, impacted, or determined by its culture. As explained previously, despite the growing usage of systems thinking terminology and buzzwords resulting from a growing number of articles, videos, and books on systems thinking, we do not believe that most organizations (or major divisions thereof) are exhibiting a culture that fosters systems thinking.

In our view, a good organizational culture has a set of *shared systems-oriented mental models* for understanding and overcoming their problems and challenges. However, as Fig. 5.14 showed, mental models are the deepest layer of the iceberg and are difficult to understand and measure. But we know that these shared mental models are *learnable* and *adaptable* because the *learning principle* from Chapter 5 states:

Learning Principle: (Common) *mental models* influencing human *decisions* about *architectures* are *learnable* and *adaptable*.

Our suggested approach to creating a culture that fosters systems thinking for organizations where linear compartmentalized mental models dominate is summarized in Fig. 9.1. The idea is to first target the mindset of leading individuals with technical and nontechnical roles by teaching them the *systems principles* and

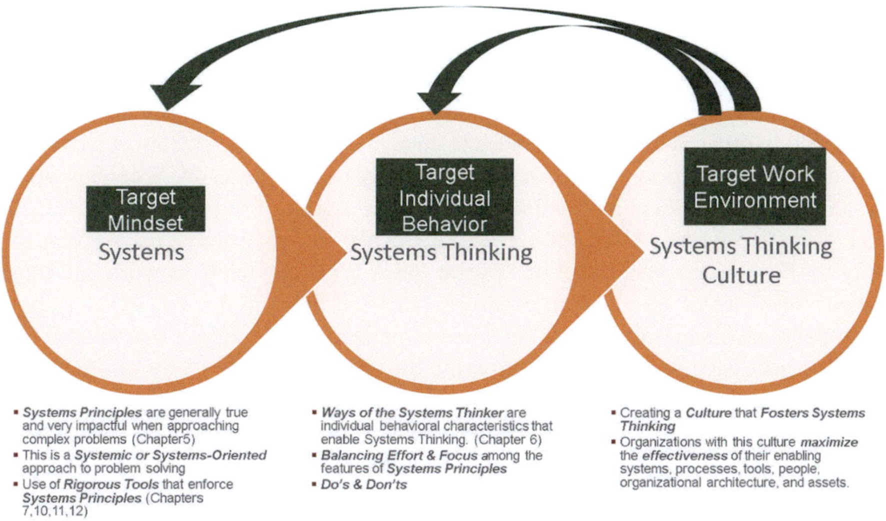

Fig. 9.1 Suggested approach to creating an organizational culture that fosters systems thinking

the *rigorous tools* that enforce them. Next is to target individual behaviors to *balance their efforts and focus* among the highlighted features of the *systems principles* as described in Chapter 7. So far, this is similar to the personal journey of the previous section and requires purposeful initiatives. The idea is to create a positive change in the proportion of systems thinking talent until a recognizable small subculture of systems-oriented mental models begins to take hold. At some point, this subculture will start regenerating and will couple with the industry's need for more systems thinking talent to naturally grow the subculture toward dominance, very similar to the reinforcing and balancing predicted by the causal loop diagram of Fig. 8.8. Once the regeneration and reinforcing effects start, the work environment will be such that it will naturally reinforce and target a more significant number of individual mindsets and behaviors.

We predict that as organizational learning begins to self-generate, businesses can optimize all aspects of their internal systems, from supply chains to project management. Employees will understand market dynamics better and be able to differentiate themselves and their teams. The organization can then leverage these skills to optimize business processes and strategic initiatives.

The problem is that since mental models are the deepest layer of the iceberg, they are also the hardest to measure directly. Establishing a hard measure of organizational progress toward a systems thinking culture is difficult. Nevertheless, the next section will explain how we may estimate and recognize such progress.

We know of several attempts to implement the approach shown in Fig. 9.1 in US organizations with mixed results. It is difficult to share the exact nature or details of the attempts without sharing the internally identifiable characteristics of the organizations. Suffice it to say that the attempts ranged from multiday training on ST philosophy to multidisciplinary team exercises solving specific complex new business challenges defined by the VP of business units. The positive results were aha moments experienced by members to better understand and appreciate their complex business challenges to increasing expenditure and utilization of modeling tools. But in other instances, resistance to change was very evident, and we received the following categories of comments: "I already have a lot to do. Worrying about systems is too technical and definitely out of the scope of my responsibilities," or "ST is just common sense, and our engineers already have it," "this is a concern only for systems engineering," or "I want lower-level engineers to work and think exactly inside their box and do exactly what I tell them to do, not wonder about what happens outside their box." In some organizations, we admit that silos are an absolute necessity so resistance to crossing the silos may be perfectly justifiable. At the same time, this type of comment can come from an organization or person that claims systems-oriented capabilities and services, in which case we can see that the ST buzzwords and terminology are not matches by actions.

We mentioned previously that it is hard to measure the impact of any change to mental models and culture since they are in the deepest layers of the problem iceberg. While we, the authors, have a range of backgrounds and industrial experiences, we do not have enough hard measures from a large pool of applications to recommend that the approach of Fig. 9.1 shall be blindly applied everywhere. Here, we believe that all organizations stand to benefit from a tailored or customized version of Fig. 9.1 if the effort balances all the concerns in the *SP*s and *SW*s for the target organization.

9.3 Measuring Progress of an Organization Toward a Systems Thinking Culture

We can recognize good systems thinking culture through the routine behaviors that exude our systems thinking principles and how these behaviors define the character of ourselves and the organization. A team or organization that has systems thinking will:

1. Purposefully try to understand the relevant complexity impacting their 3-systems. This means that teams will take the initiative to exercise their awareness and mastery of systems principles and the components discussed above. *For example, a team of city planners addressing flood risks will take a systems thinking approach by considering social, economic, and environmental factors in their discussions. They might consider the impact of flood solutions on the community systems, they will consider nature-based solutions and connections, and they will realize the interdependence of any city changes on community economies.*

   Or we might recognize this behavior in a university student striving for academic success who takes a systems thinking approach by considering academic, social, and personal health in their decisions. They might evaluate how their course workload and study regime improve their academic performance while reflecting on social commitments and recognize how maintaining physical and mental well-being supports both their learning and social lives. The term support is key here. Giving support differs from achieving a work-life balance as systems thinking moves us beyond finding balance by creating a structure that enables success in all aspects of the system.

2. Exercise systems principles in all roles and responsibilities, not just those associated with defined "systems" functions. Furthermore, teams and their members seek out and encourage systems thinking, even if those thoughts lead to tangential activities.
3. Routinely check in on the progress and activity track individual and organization growth of personal mastery. For a good systems thinking culture, this implies frequent tracking beyond annual progress reporting. *A systems thinking organization would make commitments like this: our organization conducts monthly check-ins where team members reflect on their progress with systems thinking skills and share examples of applying systems thinking principles in their work.*
4. Enable the adoption and utilization of rigorous methods of systems thinking. This adoption could include system maps during brainstorming sessions, system dynamics tools to map key business functions, and architectural modeling methods for advanced complexity management.
5. Develop initiatives to explicitly understand, map, and explore the complexity that impacts systems at all major functions or business units. *An organization would initiate workshops where teams from different business units would map out the complexities affecting their functions.*
6. Execute decisions where all the main stakeholders are present. *Systems thinking mindsets would encourage and see value in collecting and incorporating views from all affected stakeholders. Moreover, a systems thinking organization works so that key decisions are made in meetings where stakeholders are present.*
7. Limit the number of one-way informational meetings. Such meetings have the greater portion of the time allocated to one-way reporting of information and lack time for interactive feedback and sharing to include perspectives from all present.
8. Have career paths that balance short-term and long-term focused roles and responsibilities: leadership, administrative, and technical ladders that encourage multiple perspectives and interconnections between business units.
9. Balance effort and focus between the problem (need) and the solution so that running with a quick solution or trying to fit a desired outcome into decision-making is avoided. Instead, we see an active discussion of the problem, solutions, and system elements that define the behaviors that have resulted in or contributed to the problems.
10. Initiate community-oriented talent flow programs where industry players join forces with academia to overcome the shortage of systems-oriented skills. The creation of the Colorado State University (CSU) Systems Engineering Program

is a successful example of this suggestion. Several prominent employers who deliver complex products and services teamed up with CSU faculty to design and kick-start this program in 2009. It has now grown to 15 full-time faculty, factors more associated faculty from other departments, and 500+ graduate students enrolled in certificate, master, and doctoral programs.

11. Have tailor-able internal iterative processes that align with systems-oriented industry standards. *For instance, a software company implements a flexible agile framework, aligning it with systems-oriented industry standards like the INCOSE Scaled Agile Framework® (SAFe) practices. The company tailors its SAFe iterative sprints by integrating feedback loops to support systems thinking principles: the* ***holism principle*** *(consider elements and interconnections collectively), the* ***architecture principle*** *(adapt the software application's structure, purpose, and behavior of the system to recognize learning), and the* ***3-systems principle*** *(account for context (market trends, user requirements, technological changes), the system of system (the software), and enabling systems (tools, platforms, infrastructure)). The software company would also support the* ***boundaries, evolution, learning, and human principles*** *by redefining system boundaries as needed, reviewing mental models, and recognizing the role of humans in the system.*
12. Enable interactive, live, dynamic discussions and dialogue between team members. Teams seek out discussion rather than deferring or delegating activities.

9.4 Reflection on Chapter 9 Material

In this chapter, we suggested a path for personal mastery of systems thinking principles and rigorous tools (like system dynamics) that help enforce their application to real-world problems. We discussed how rigorous tools like system dynamics can encourage our learning and growth so we can better embrace systems thinking as a core philosophy for understanding and approaching real-world problems. We then suggested a three-step incremental and iterative approach to creating an organizational culture that fosters systems thinking in all their significant activities (shown in Fig. 9.1). We then discussed how we might measure an organization's progress toward a culture that fosters systems thinking. After reading this chapter, we hope the reader is convinced that there is a practical path toward obtaining a systems thinking mindset. As we read on to Part III, to learn more about applying rigorous systems thinking tools, we may slowly realize that we have already started our journey toward achieving a systems thinking mindset!

Next, we begin a deeper dive into rigorous frameworks, languages, and tools that enable, enforce, and implement systems thinking in Part III.

Part III
Rigorous Tools for Systems Thinking

The key to success isn't just thinking about what we are doing but doing something about what we are thinking—*Peter Senge*

Chapter 10
Real-World Applications of Rigorous ST Methods

Abstract A way we can evaluate an organization's path toward adopting a systems thinking culture is to measure the penetration of rigorous methods that enforce it. Our observation is that despite the increasing number of challenges many organizations face, adopting rigorous frameworks, tools, and methods is very low across engineering, administrative, sales, and marketing functions. Since most fresh young talent will not usually be exposed to rigorous systems thinking frameworks, languages, and tools when they start their careers, we give several real-world example applications here.

10.1 Rescuing Runaway Projects

A significant proportion of projects in best-in-class companies that have completed similar projects successfully and employ experienced and certified professionals end up costing much more and taking much more time than initially planned. For example, it may take an aircraft manufacturer 4 years and on average 55 people per year to complete a new aircraft project. Yet, another aircraft project that before starting appears to have very similar objectives at the same company using the same team and processes will take 11 years to complete the project.

Another example is the VA Hospital replacement in Denver, initially estimated to cost $328 million with a planned opening in 2013. When the opening ceremony finally occurred in 2018, only half the original plan was completed, and the cost had ballooned to $1.73 billion. Since only half was completed, the original aging VA facility had to stay operational at a cost of $350 million per year. Media (NBC News), experts, and official government reports (Wise) offered only a vague explanation that the issues "are primarily the result of poor business decisions, inexperience with the type of contract used, and mismanagement by VA senior leaders". We find examples of runaway projects everywhere: California High-Speed Rail, Webb space telescope, Joint Strike Fighter, etc.

Understanding the causes of a runaway project requires knowing that a project outcome is a complex emergent behavior resulting from the interaction of feedback loops like the interaction between the *rework cycle*, *staffing loops*, and loops that

K. Eftekhari Shahroudi et al., *Practical Systems Thinking*,
https://doi.org/10.1007/978-3-031-89394-0_10

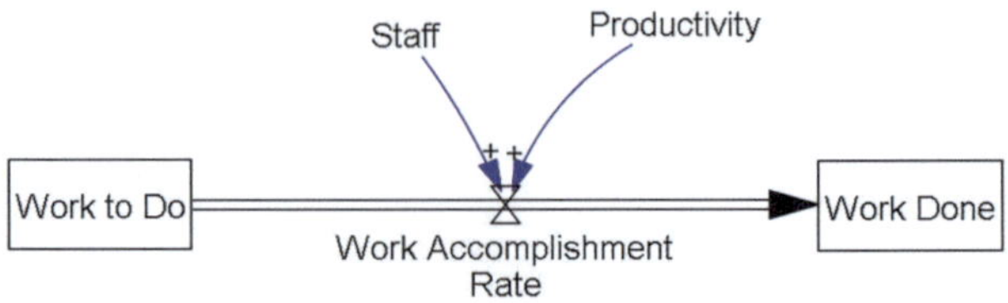

Fig. 10.1 Classical or simple view of work accomplishment assumes that work accomplishment rate equals work assignment rate (assigned staff x productivity)

enforce project management decisions such as *schedule priority*. Using system dynamics modeling provides the necessary insight into the causes of this complex emergence. For example, a system dynamics model shows that for the project modeled, the "time to gain experience" has minimal impact on project "cost" and "time to complete" relative to "normal quality." It also shows that an aggressive staffing policy mid-project to finish the project on time leads to a more costly and longer duration project, which is the opposite of the intended outcome of adding staff. Those who do not have systems thinking skills typically follow their classical "boxed-in" or linear thinking approach to identify events or outcomes that are readily observable (i.e., the tip of the iceberg) and link them to a cause that is nearby in time and location. Linear thinking typically cannot identify the actual mechanism of policy, mental models, or the structure (deeper layers of the iceberg) that drives the complex emergent behavior runaway projects. System dynamics (SD) [9] is among the most useful rigorous systems thinking frameworks, languages, and tools for understanding complex emergence where the cause is not a simple or linear combination of known or easily observable effects.

The SD diagram in Fig. 10.1 represents the traditional or straightforward or ideal mental model of how work gets accomplished in a project. This implies that work gets accomplished at a rate that equals how work is assigned: assigned staff x productivity. As work gets accomplished, tasks flow from the *work to do* pile into the *work done* pile at a scheduled rate, like water flowing from one bathtub to another through a controllable water valve. Here, the project manager only worries about scheduling the tasks and assigning them to people using a cost and schedule buffer of, say, 10% to cover for uncertainties. So, this simple mental model of a project accounts for minor issues but cannot explain emergent runaway project behavior.

Figure 10.2 gives a better mental model of project dynamics that accounts for rework generation. Here, we recognize that as we work, we make errors or changes that ripple through the systems, eventually creating more work called *rework*. But *rework* typically takes a while to discover, so it piles up in *undiscovered rework* pile before it is discovered and put back into the *work to do* pile. The parameter that controls the proportion of *undiscovered work* that flows back into the *work to do* pile versus the proportion that gets accomplished is *quality*. However, quality itself is not constant because rework has a reinforcing feedback effect on quality.

Figure 10.2 also compares the ideal versus rework mental model responses for a *quality* of 75%. Quality quickly drops due to the action of the Q^2 reinforcing loop

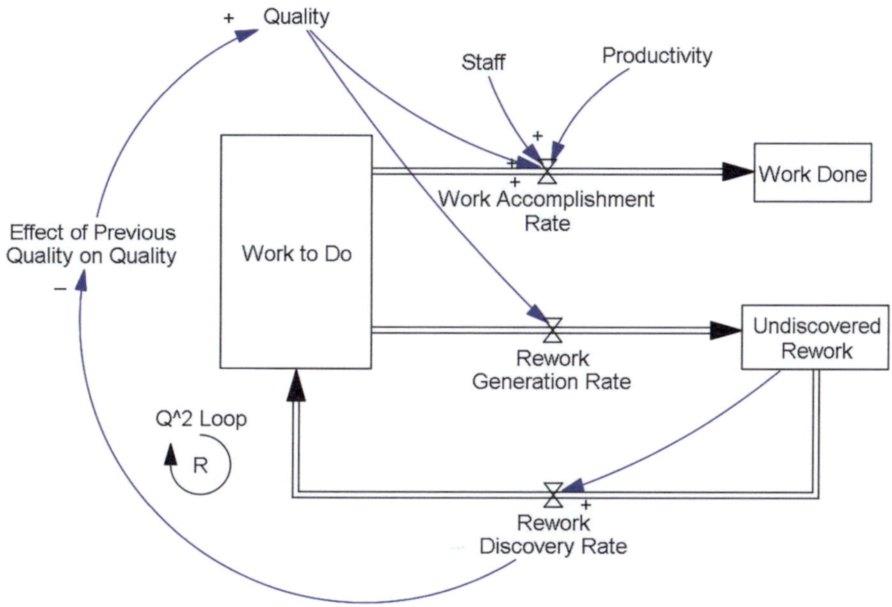

Fig. 10.2 Response of the *rework cycle* mental model with Q^2 reinforcing loop versus *traditional*

and then slowly recovers over time. Generating rework has the impact that both the *work done* and *work to do* move slower, so the project takes longer. Now, the project manager (PM) is entirely unaware of issues until the *rework discovery rate* becomes significant before halfway through the project. At this point, the PM increases the assigned staff to increase the *work accomplishment rate*. The difference in the area under the *staff level* curves is closely related to the cost difference that the two mental models of the project can predict. Hence, the improved mental model and representation of project dynamics in Fig. 10.2 can better predict the extra cost and duration of the project that emerges from rework.

Yet, a few more loops are important to improve the mental model so that we can understand or predict the emergent behavior of projects better. Figure 10.3 shows additional feedback loops that impact *quality*, *staff*, and *productivity*. These additional loops capture the monitoring, staffing decisions, and the time it takes for new staff to learn to deliver higher-quality work more efficiently. Figures 10.4 and 10.5 show the impact of adding these loops and compare the dynamic response of the *classical*, *rework*, and *rework with staffing* models. Note that the latter can predict the runaway project dynamics described earlier, where the project is significantly more costly and delayed than initially anticipated.

The *3-systems* principle teaches us that systems always come in (at least) 3s. So, even if we aim to improve the efficiency of work/rework cycle throughput (the system of interest or SOI), we need to consider all the relevant enabling and context systems, their interconnections, and interactions with the SOI. Figure 10.6 shows a

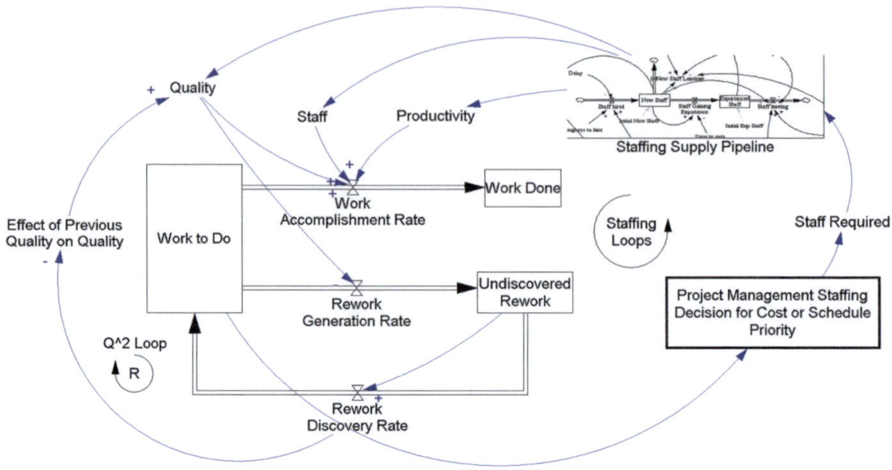

Fig. 10.3 Rework cycle with staffing loops that capture PM decisions for *schedule priority*

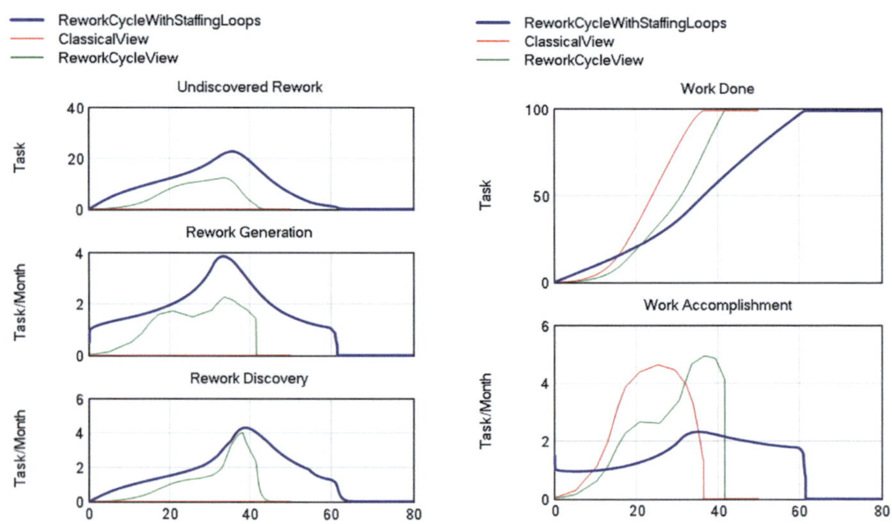

(a) *Undiscovered Rework* and its causes (b) *Work Done* and its cause

Fig. 10.4 Comparison of *undiscovered rework* and *work done* responses of three ways to model project dynamics

system dynamics model programmed Vensim software that has the hiring pipeline (enabling system), project tracking and staffing decisions (the context system), and their interconnections with the work/rework cycle (the SOI) modeled.

Once we have a validated SD model that includes the 3-systems and all the relevant interactions among them, we can answer questions that ordinarily confuse linear thinking project managers. For example, which of the following scenarios

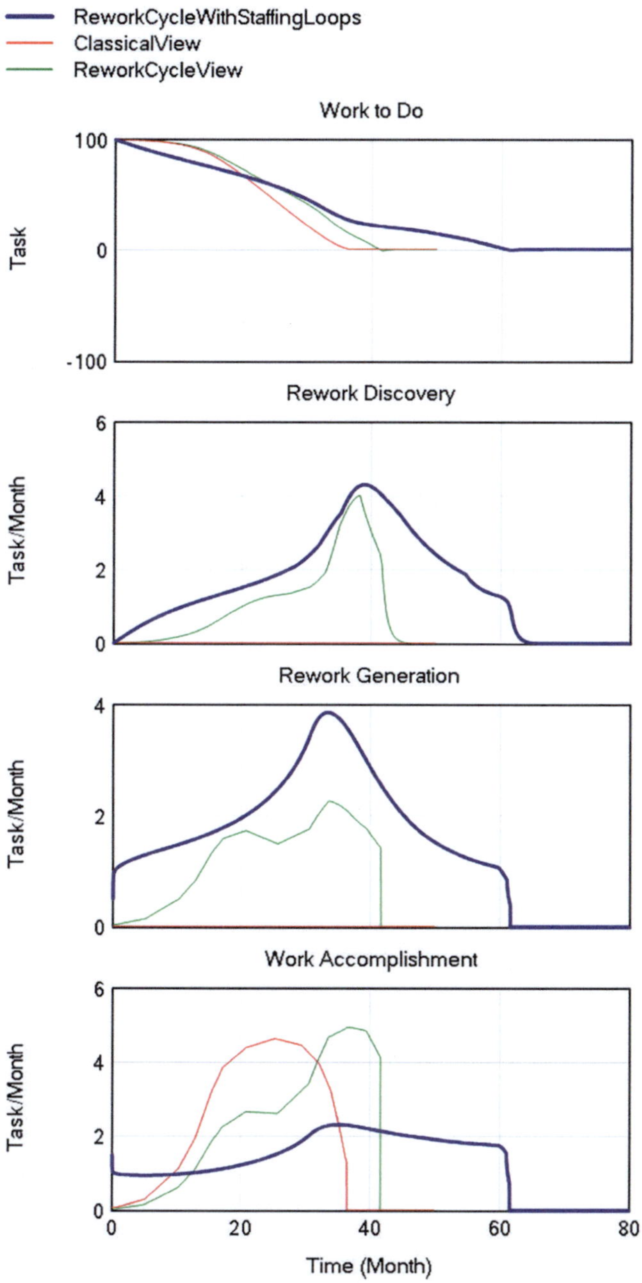

Fig. 10.5 Comparison of *work to do* response of three ways to think about and model project dynamics: classical, rework, and rework with staffing loops

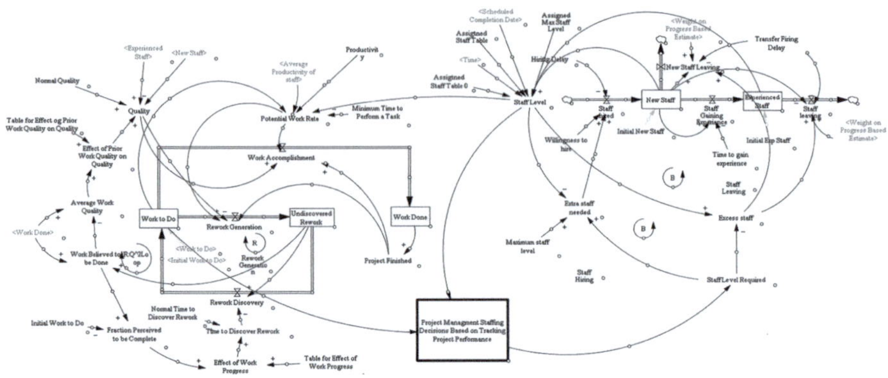

Fig. 10.6 SD model programmed in Vensim software containing the SoI (work/rework cycle), context system (scheduling and staffing decisions), and enabling system (hiring/staffing pipeline)

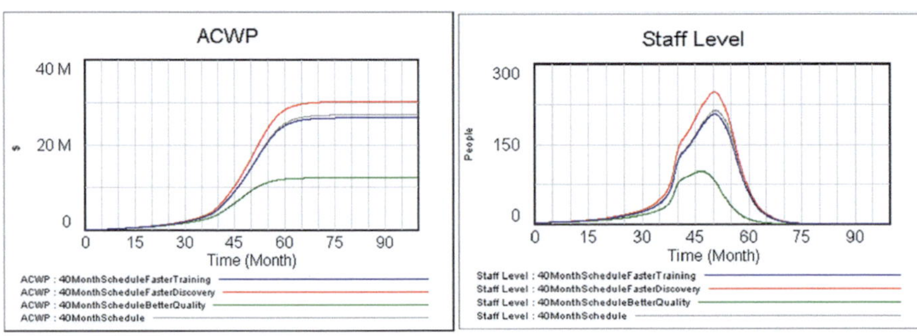

Fig. 10.7 Comparison of cost (ACWP) and staffing impact of faster training, faster rework discovery, and better quality using the Vensim SD model of Fig. 10.6

will significantly impact cost while improving the completion date: better quality, faster time to gain experience, or faster time to discover rework?

Figure 10.7, computed using the Vensim SD model of Fig. 10.6, can answer this question. It shows that faster *time to gain experience* has minimal impact, while improvement in the *quality* significantly reduces the cost while completing the project sooner. A counterintuitive result is that for this particular project, the faster discovery of rework, which is normally considered a process improvement, actually increases cost or makes matters worse! So, spending money to locally improve a part of the SOI leads to making the whole system worse because of the way the SOI interacts with the context and enabling systems! This may be counterintuitive to a linear compartmentalized thinking project manager who believes improving a part will naturally lead to improving the whole. This shows the power of system dynamics.

At this point, we can see that system dynamics allows us to understand how the *structure* (the loops with stocks and flows) and the *behavior* (responses in

Figs. 10.5 and 10.7) and *purpose* (cost and schedule targets) influence each other. In other words, system dynamics also enables us to rigorously apply the *architecture principle* that states:

Architecture Principle: The *purpose*, *structure*, and *behavior* of systems *influence* each other.

The validated SD model of Fig. 10.7 can be expanded to answer more real-world complex scenarios. For example, is the sequential execution of a particular program (or group) of projects better than parallel or agile execution? What are the unwanted side effects of schedule priority compared to cost and scope priority for a particular project? Under what conditions is adding staff to a running project helpful or hurtful? Answering these questions requires a deeper understanding of the interacting loops within and among the three systems. System dynamics is one of the most essential and valuable methods in model-based systems thinking (MBST). MBST provides a deeper understanding of complex real-world problems and integrates several other valuable modeling methods as described in more detail in Chapter 11.

At this point, we can see that system dynamics also allows us to understand how complex behavior *emerges* from the feedback loops and other interactions between the work/rework cycle, staffing pipeline, and project management, in addition to how humans (staff gaining experience, PM staffing policies and decisions). In other words, system dynamics also enables us to rigorously apply the *emergence* and *human* principles to improve our understanding because they state (from Chapter 5):

Emergence Principle: *Interactions*(including feedback) within and between the 3-systems cause (sometimes unwanted) *emergent behaviors*.

Human Principle: *Humans* impact and are impacted by the (wanted/unwanted) *emergence* of *all* reachable systems.

Teaching the system dynamics modeling method at CSU revealed that most students do not naturally possess all the necessary skills for it regardless of their disciplinary affiliation and domains of interest. They require training to become proficient in systems thinking philosophy, rigorous methods, and system dynamics framework, language, and tools such as Vensim, Stella, or iSight. The initial objective of SD modeling is to find a structure (i.e., an executable SD model) that can reproduce some known or measured pattern of behavior. Validating the model by reproducing the known pattern of behavior gives us the confidence to go further and explore sensitivities, scenarios, and policies to improve the response. Following the system dynamics method enables and forces us to rigorously apply the systems principles. The final goal is to develop a better mental model of real-world problems and to create policies that drive better behavior while predicting their sensitivity and unintended consequences. The standard text of John Sterman [9] is an excellent source for finding SD examples of a wide range of real-world problems. At CSU, we had to adapt and extend the SD method of Sterman to clearly distinguish slow/fast learning loops and bring in other modeling methods that have proven their benefit in understanding and dealing with complex real-world problems, such as AI/ML, agent-based modeling, architectural modeling, etc. We call the extended method the model-based systems thinking method, which is the topic of Chapter 11.

10.2 Elevating Wildfire Response

Wildfires have become a global concern that has grown with each new incident worldwide in recent years [7, 10] with most recent occurrence in Los Angeles, California, in January 2025 and Hawaii in August 2023. For example, wildfires in the western USA can emit more carbon dioxide in just one week than all cars in an entire year [6]. However, the consequences of wildfires are not limited to any location. Early detection and effective communication become crucial to minimize their adverse impact on wildlife, humans, and the environment.

One of the most compelling solutions is turning out to be unmanned aerial vehicles (UAVs). Interestingly, not all UAVs are suitable or efficient for wildfire detection and communication, as most UAVs are designed for universal activities. Classical general-purpose UAVs cannot detect wildfires early due to their deployment strategies.

The priority for wildfire detection should lean toward specialized sensors for detecting heat and smoke, real-time data relays in adverse conditions, and potentially rapid response tools. Therefore, a UAV explicitly designed for wildfires offers the best combination of cost-efficiency and performance, as these UAVs can save on fuel efficiency and extend the range of the aircraft by taking advantage of both aircraft and wildfire parameters.

Given the locations of wildfires, it is apparent that standard runways might not always be accessible in areas prone to wildfires. Our UAVs are therefore equipped for vertical lifts and are taken near locations prone to wildfires using trucks as mobile base stations. Each UAV maintains communication with its respective mobile base station.

It is also possible for these UAVs to communicate with central base stations as needed for general operations. The UAVs can form a mesh communications network while operating in the sky. The UAVs can intercommunicate using the mesh network in the unforeseeable event of satellite communication disturbances due to terrain complexities. This provides redundancies and strengthens the overall system. The response time to fire departments is also a factor that must be considered for this UAV design. In the event of a city, town, or county emergency system failure, sirens may be needed to communicate rapid-spread wildfires to human habitats.

The systems principles are closely applied in order to evolve from traditional flight control architecture to novel surfaceless flight control (SFC) fan-in-wings (FIW) concepts that reduced the weight of the aircraft and effectively translated into increased range and/or increased payload.

Thinking in 3-Systems The mental model is to design an UAV for wildfire to maximize range while minimizing production and manufacturing costs, making it accessible to local wildfire fighting organizations.

3-Systems Principle: *systems* come in (at least) *3s*: *context systems*, *systems of interest (SoI)*, and *enabling systems*.

Through a series of systemic steps of the MBST framework, we have discovered a novel and efficient architecture, customizing the UAV airfoil design and flight

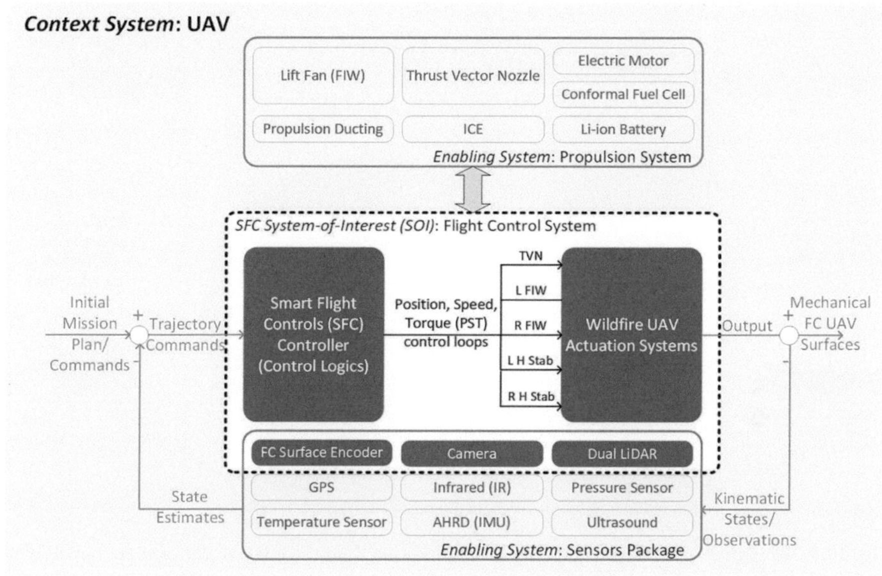

Fig. 10.8 Applying 3-systems principle—flight control system as SOI [2]

controls with fan-in-wings (FIW) to address the specific needs of wildfire detection and communication shown in Fig. 10.8 identifying three types of systems: SFC system-of-interest (SOI), enabling systems (such as propulsion system, sensor package), and entire context system (the UAV in this case).

This architecture can significantly reduce the costs of manufacturing and operating UAVs while drastically improving the performance of wildfire detection and monitoring. Our UAV design includes a game-changing surfaceless flight control (SFC) system, which radically shifts away from the existing UAV architecture and redefines the overall UAV operation to focus on wildfire detection [6]. By omitting traditional and general-purpose flight control aspects, we enhance the UAV's aerodynamic capability and reduce its weight [2].

Using model-based systems engineering (MBSE) as a rigorous tool (Cameo) allows for systemic modeling of complex architecture such as a flight control system, UAV subsystems (e.g., propulsion, sensor package), and enabling systems. By applying systems principles (3-systems principle in particular), MBSE helps improve system architecture. The interaction between these tools and the principles of the systems ensures that we develop an effective flight control system, which leads to better performance and enhanced design at various stages of development.

Holistic Thinking

Our goal expanded beyond simply meeting the pitch, roll, and yaw control requirements or simply reducing weight. Instead, we adopted a holistic systems approach to comprehensively evaluate the overall system.

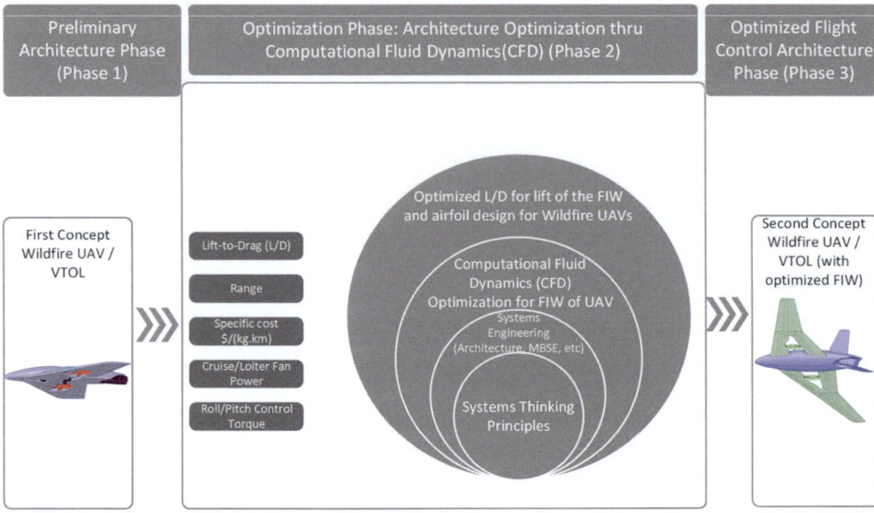

Fig. 10.9 Applying systems principles—from traditional flight control concept to novel FIW concept [3]

Holism principle: (complex) *systems* are (much) *more* than the *sum* of their *parts—elements*, *interconnects*, and *purpose*.

As illustrated in Fig. 10.9, we identified opportunities to eliminate certain flight control components from the preliminary design (left side of the figure) and implemented a new design (right side of the figure) that significantly reduced weight while enhancing system performance.

In addition, we optimized the use of flight control fan-in-wings (FIWs) for pitch, roll, and yaw functions. By strategically positioning the FIWs to maximize lift, we achieved improved aerodynamic efficiency. This repositioning also allowed us to reduce the wing surface area, further decreasing weight without compromising performance. Furthermore, we introduced a gap at the wing root, an area less sensitive to aerodynamic contributions, which positively influenced the lift-to-drag (L/D) ratio, further enhancing overall system performance.

Using CFD (Ansys CFX) as a rigorous tool, the design process becomes more rigorous, allowing us to optimize airfoil and wing design and essentially improve the overall efficiency of the UAV. Combining holistic thinking with aerodynamics and CFD modeling ensures an integrated approach to system design and performance optimization.

Evolutionary Thinking The UAV design evolution reflects a move toward a lighter, more mission-oriented architecture based on wildfires. By improving UAV adaptability and minimizing its weight, the design evolved from traditional flight control architectures to an innovative surface-lift flight control concept with a fan-in-wing design. This approach not only reduced the aircraft's weight but also effectively increased its range and payload capacity.

Evolution principle: systems have *life cycles* and they *evolve*.

In Fig. 10.9, the principles of evolution are evident. We reviewed the initial preliminary design to identify the changes required to refine and advance the concept. The key evolutionary breakthrough was the redesign of the flight control system and its integration with the fan-in-wing concept. This redesign provided an opportunity to collaboratively develop and validate a computational fluid dynamics (CFD) model, addressing gap wings with FIW functionality, FIW locations, and aerodynamic objectives.

Using a rigorous tool like CFD, simulating scenarios like virtual wind tunnels, helps to model and evaluate the aerodynamic design of the UAV. By applying systems thinking principles (holism principle, evolution principle), such as feedback loops and optimization, we can evolve the preliminary architecture of the UAV's flight control system. Through this process, the design evolves to become more efficient and the flight controls are optimized. The combination of evolution and holistic thinking ensures that the UAV's design is optimized across various stages of development.

When we step back and observe the entire complex problem of wildfire, it becomes evident that the unique needs of wildfire detection and communication call for UAVs designed specifically for that use. Our game-changer design includes an airfoil and flight control design to minimize weight and cost. It specializes in early detection, extended loitering range, instantaneous monitoring, accessibility to rugged terrains, extended operation time, extended communication availability, data collection, safety, and rapid deployment.

10.3 The Same-Day Missed Appointment Problem

When we schedule appointments at most doctor's offices, that time slot is reserved for us and only for us. It is not like booking a ticket for a flight where there is a chance that the airline has double-booked and someone else could take our spot. Nope, that time slot is for reserved just for us and they expect us to show up. A lot goes into booking that appointment too. Schedulers at the office spend time putting our personal information into a system and working with us to find the right time slot. They may even check our insurance eligibility and tell us how much of the cost our insurance carrier will cover and how much of the cost we will be responsible for. Then, on the day of our appointment, someone in the facility will make sure the room is clean and ready for our visit. They gather up any materials needed for our visit and wait in anticipation for us to show up. Everyone involved in setting up everything for our visit is paid by the organization they work for. In healthcare, the organization only receives payment from insurance companies and the patients themselves if they are seen. Patients never show up to pay us for the time spent setting up their appointments if they were not seen. Therefore, when we miss our appointments due to a variety of reasons, not only does the organization never

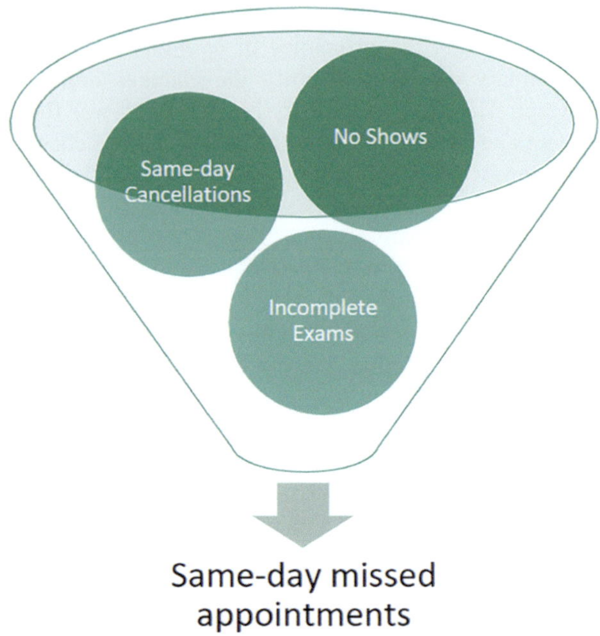

Fig. 10.10 Everything that falls into the category of same-day missed appointment

recoup the cost of what they had to pay their staff to prepare for the appointment, but they are also put in a position where often they cannot fill the vacated time slot at the last minute. This can be very frustrating for doctor offices with a long list of patients trying to be seen. This is another example of the *human principle*. Not only did the patient, for whom the organization spent a lot of time scheduling and preparing, not show up, but another patient who would have liked to have been seen sooner was left hanging. There are also cases where as patients we may show up and then cannot be seen. We may be unable to pay the out-of-pocket cost, get claustrophobic during our MRI (magnetic resonance imaging), or refuse services altogether. These are just a few examples of the many reasons that we can miss an appointment on the same day. Patients who are no-shows and patients who are unable to be seen on the same day all fall under the category of same-day missed appointments (see Fig. 10.10).

Since missed appointments occur for many reasons, there is no simple fix to the problem. This is a complex problem. Two standard solutions healthcare providers have implemented are automated reminder calls to proactively address the problem and no-show fees to address the issue reactively. These local solutions may provide some initial benefits but do not fix the problem in the long term. Again, patients miss appointments on the same day for many reasons, and appointment reminders only address the "forgot" reason. No-show fees are established to help the provider recover some of the lost revenue but do not address the reason for the no-show, and the providers have reported mixed results in implementing this "solution" [8]. The *emergence principle* reminds us that just addressing one cause without analyzing all feedback loops is why these solutions have had limited effectiveness.

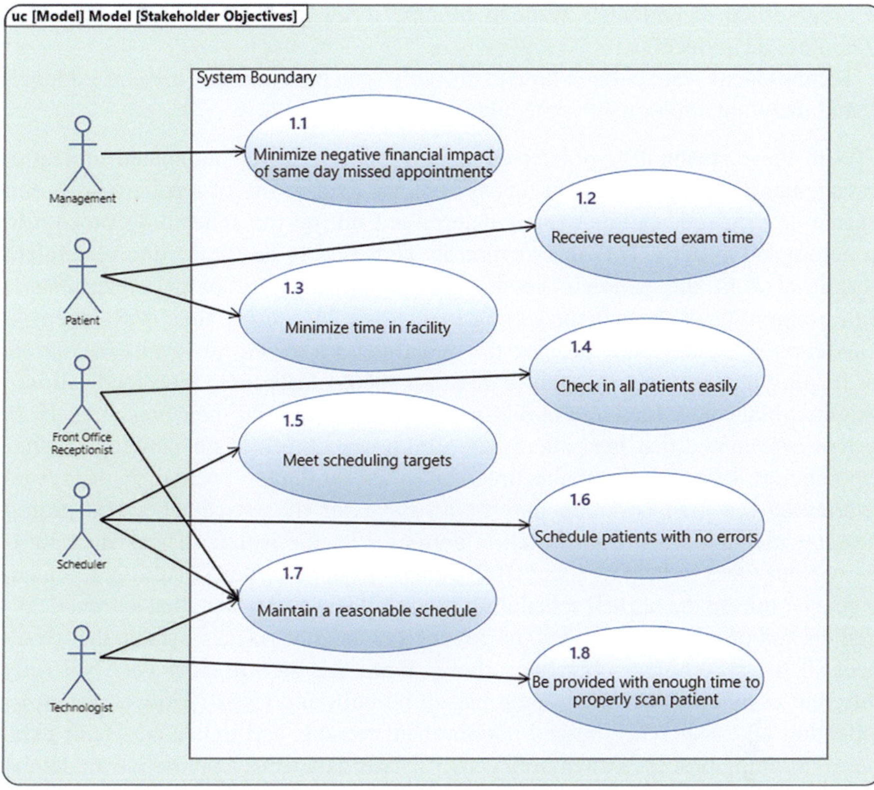

Fig. 10.11 Mission level use case objectives for a system that automates utilization management

An outpatient imaging center on the central coast of California that has been studying this issue for years concluded that this complex problem requires a more sophisticated solution than had been attempted in the past. Sustained solutions require understanding different viewpoints while remaining focused on the big picture. The solution to this problem had to consider the impact on the whole forest while also understanding the impact on each tree, i.e., the *3-systems principle*. The first thing that needed to be done was a full understanding of the primary stakeholders' needs as seen in the use case diagram in Fig. 10.11. Through interviews and direct observation, the team found the following:

- Patients want to be seen at their requested exam time and want to minimize their time in the facility.
- The management team wants to reduce costs by reducing the same-day missed appointment rate and filling all available time slots.
- Schedulers want an easy error-free system for scheduling patients and want a process that allows them to meet their scheduling targets (the number of exams each scheduler needs to schedule every day).

- Front office receptionists want to be able to easily check in all patients and address any concerns.
- Technologists want enough time to properly scan patients, take breaks and lunch, and maintain a reasonable schedule.

With these stakeholder objectives in mind, a system for automated utilization management was envisioned. In this system, the probability of a patient becoming a same-day missed appointment is determined during the scheduling process by an automated system. This information is then used to customize the scheduler's workflow to fit the patient's needs. For example, if the model determines a high probability of the patient getting claustrophobic during their MRI, then the scheduling system would prompt the scheduler to review all available options for helping a patient get through their exam successfully—sedation medication, a weighted blanket, a friend or family member ready to hold their hand, etc. If the system determined that the patient had a high probability of no showing for their appointment, then that particular time slot could be double-booked. If the system determined that transportation may be an issue for the patient, it would prompt the scheduler to review transportation options with the patient. There are a lot of options out there to help us successfully keep our appointments, but it is too many to go over during the normal scheduling process. Hence, a system that automatically predicts and proactively supports the patient and the business is a system that would meet all the stakeholder objectives above. Does this system exist yet? Not fully. Different components of the system had to be built and tested. First, a system for collecting and displaying missed appointment reasons and trends was built using the A3 thinking process shown in Fig. 10.12. This data feeds a predictive model that must be manually run daily but helps identify a targeted call confirm list. A system was also built to provide patients with their estimated wait time once they arrive at

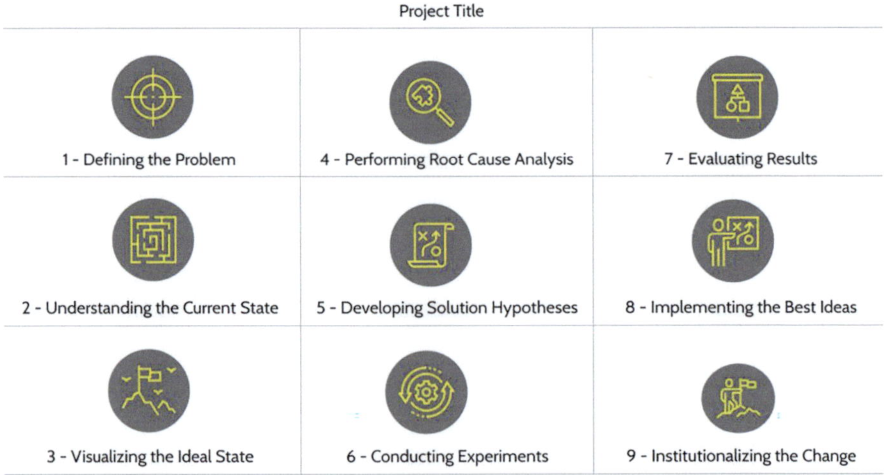

Fig. 10.12 An A3 of the process of creating a same-day missed dashboard

the facility. More pieces of the system need to be fully developed but a road map has been developed to realize the full vision, and some stops along the way have been successfully checked off!

10.3.1 How to Apply the Model-Based System Architecture Process

The goal for this system architecture is to build a system that optimizes utilization without dependency on modifying patient behavior. The Model-Based System Architecture Process (MBSAP) provided a comprehensive and visually understandable framework for system development in an industry unfamiliar with systems engineering methods. Some examples of the components of MBSAP that were used to develop the architecture of this system in outpatient imaging are outlined in the following paragraphs.

An initial domain diagram for the automated utilization management systems shows the three primary domains—an exam status system, a radiology information system (RIS), and a predictive model. Each domain has its own unique parts and user interactions.

An enterprise domain diagram shows how this system could fit into the greater healthcare system. An automated utilization management system would be even more effective if the overall healthcare enterprise could be changed to provide a central information repository. A central information repository would be used to collect and distribute patient orders and information. There are security issues and HIPAA considerations. Still, the benefits of this type of system (improved healthcare accessibility for the whole community) potentially outweigh the cost of managing the security considerations.

A reference architecture diagram shows how the architecture could be used in any business that schedules appointments, has regular cancellations, is in high demand, and desires to proactively manage its utilization.

Use case diagrams were created for staff functions. For example, front office receptionists will need to be signaled if a patient does not check in by their scheduled table time and reach out to the patient to obtain a reason for missing their appointment. This information could be fed to the predictive model to improve the accuracy of workflow directives.

A sequence diagram with timing shows how the requirement to not exceed current scheduling cycle times can be met.

An activity diagram shows the process for modifying scheduler workflows if the patient's cancellation probability is high and an actionable cancellation reason has been identified. It also shows the process for the scenario in which the patient's cancellation probability is high and an actionable cancellation reason cannot be identified. In this scenario, the predictive model will send the RIS a trigger to allow the exam time to be double-booked, but the scheduler's workflow will not change.

10.4 Security by Design in Complex Cyber-Physical System Development Teams

Cyber-physical systems (CPSs) are increasingly prevalent and interconnected in our modern world. CPSs include airplanes and vehicles and other complex systems that combine software and hardware and rely on both for accomplishing their primary functions. While CPSs are not new, they have shifted from being primarily mechanically driven functionality to software-controlled and interconnected via cyber. This has increased many positive features such as usability, safety, reliability, and performance, but it comes with some trade-offs in increasing the vulnerability space to cyber-threats. Present-day cyber-physical systems (CPSs) are designed and fielded by teams of engineers and program managers, who often pay inadequate attention to cybersecurity. The structure of the CPS design team requires investment in systems security and systems thinking to create a culture that can achieve "security by design". Modern CPSs are increasingly interconnected in the industrial Internet of Things (IIOT), with many not applying a systems approach to understanding the vulnerabilities induced by this functionality. Cyber-attackers' mental model holistically views the entire CPS, finding vulnerabilities in the limited IT security and emergent system properties that need to be exploited. Therefore, there is a need for a systematic approach to incorporate "security by design" for the entire CPS, and there are lessons we can learn on the value of systems thinking as applied to this problem set.

Cybersecurity of CPS is a complex problem as the attacks are a complex chain of events and emerge as a complex dynamic process. Vulnerabilities are often not known until they are exploited, causing zero-day attacks. Additionally, attackers view the CPS holistically and evolve with new techniques and tactics to exploit vulnerabilities. Often, the CPS design team is not equipped with the mental model to look at these vulnerabilities and possible attacks as cyber-attacks fall outside the area of their core responsibility. Attackers discover new techniques to exploit the system, but many current CPS design teams lack the expertise to address these issues during the CPS design.

A complex cyber-physical system (i.e., airplane, satellite, truck) incorporates several electrical, mechanical, controls, software, and communications subsystems. Many of these subsystems rely on software, Internet, over-the-air, and remote electronic transmissions to perform functionality. Too often, security design experts are not even included in the design team. Many companies and CPS design teams fail to systematically incorporate "security by design" for the entire system. Systems thinking encourages a holistic perspective to problem-solving beyond traditional reductionist methods that have failed to solve some of the world's most complex problems. Complicated problems can often be solved by using this reductionist method of decomposing the problem into smaller parts. This approach fails in CPS as they are more than just complicated; they are interactively complex. A CPS is more than a sum of its parts; it requires a systematic approach that considers the system holistically. This is a key tenet of systems thinking. The traditional CPS design teams have several shortcomings. First, the requirements and design teams do not cover security aspects. Second, the CPS design team relies on IT personnel for security. Third, CPS design teams do not study emergent behavior. Such behavior often generates vulnerabilities that can be exploited.

The following are the shortcomings of traditional CPS design with respect to the various systems principles:

- **Holism**: The CPS design team does not view cybersecurity holistically. On the contrary, attackers view CPS holistically to exploit the vulnerabilities.
- **Evolution**: Current CPS does not accommodate system evolution, but the attacks evolve.
- **Emergence**: Security is an emergent property, and often vulnerabilities result from emergent behavior that results from the interactions between elements of the system and the environment.
- **Feedback:** Vulnerabilities emerge as unwanted system behaviors often not apparent in early design but emerge from feedback loops with delays.

A reductionist approach to solving this problem has not been successful (US Government Accountability Office) (Checkoway et al. [4]). Securing the individual Internet of Things (IoT) devices within a CPS with current IT security practices has proven unsuccessful as it fails to consider the system's emergent properties. Ironically, those who seek to exploit CPS use a holistic approach to determine vulnerabilities in the system based on the complexity of interactions between the system elements and the environment. The CPS design teams must employ a holistic systems approach utilizing systems thinking principles to protect against unknown cyber-exploits. Threat and exploit tactics continuously evolve. Thus, the design must approach security proactively and invest significant effort into "security by design".

10.4.1 How to Use the Iceberg Model Effectively

The iceberg model is a simple tool for gaining a deeper understanding of a problem by seeing beyond the apparent trigger event(s) (e.g., CPS outages caused by successful cyber-attacks). The iceberg model guides an understanding of the buried root cause by identifying patterns of behavior associated with the event, uncovering the underlying system structure responsible for creating those patterns, and uncovering the underlying mental model—in multiple levels of detail—to understand better the underlying structures which lead to the undesired, hazardous, or harmful event. Figure. 10.13 is an iceberg model showing traditional CPS security design team failings.

- **Cybersecurity Events**: "Events" are observable unwanted, undesired, or potentially hazardous occurrences. They do not necessarily have to be malicious or harmful but merely indicate a problem that needs further investigation. As indicated by the "iceberg" analogy, the problem is often much larger (and more

Fig. 10.13 Iceberg model for CPS design team cybersecurity challenges

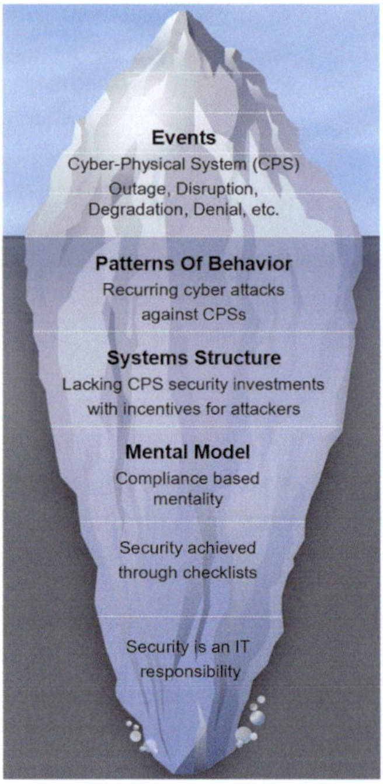

complex) than is initially observable, and most of the challenge is hidden from plain sight. In the case of CPS design teams, the "event" is witnessed through successful cyber-attacks that occur against CPSs of all scope, scale, and industry. Recent world events demonstrate the ability of malicious actors to continue to exploit not only IT networks but critical CPS devices and CPS-enabled SoS [11]. Too often, CPS devices fall victim to cyber-attacks, resulting in unwanted disruption, delays, denial of services, or worse; thus, the analysis continues.

- **Patterns in Security**: At this level of the root cause analysis, patterns of behavior are reoccurring unwanted activities. In our analysis, patterns represent shortcomings of traditional CPS designs and teams that result in vulnerabilities and cyber-attacks. For example, system specifications that state "the system shall be secure" fail to understand the importance of cybersecurity for modern connected systems. Poor CPS security patterns are evidenced by the increasing number of vulnerable CPS devices being developed and the widespread exploitation of everything from simple home-monitoring video cameras to advanced automobiles in the popular media [5]. Observed patterns within CPS design teams may include pressure to bring new devices to market quickly, an over-narrow focus on desired functionality without sufficient concern for security requirements and/or privacy considerations; treating security as an afterthought; reliance on IT personnel to address system-level security issues; trusting in IT-based checklist security approaches; the ability to patch CPS devices as required in the future to fix problems; a limited understanding or refinement of stakeholder needs for safety or security. Perhaps most importantly, key systems thinking tenets of holism and emergence are often not considered. However, attackers view CPS holistically and seek out novel and unique ways to exploit any vulnerabilities, not just the ones the designers plan for. Historically, ignoring emergent behaviors can result in exploitable vulnerabilities that malicious actors can exploit [4]. Thus, security is an emergent property where vulnerabilities often result from unanticipated behaviors resulting from interactions between and within elements and users of the system of interest (SoI). These interactions must be modeled and studied, allowing the CPS design team to understand the CPS security problem better.

- **Systems Structure of Cybersecurity**: The totality of the system's structure can be wide in scope. We will keep our focus on the design teams charged with securing new CPS systems. Typical structural concerns include CPS companies and their design teams, national policies and regulations, external operational context and existing infrastructure, social structures, and personal motivations, all contributing to CPS vulnerabilities and attacks. In most cases, the structural incentive for CPS attacks has largely gone unmatched by regulatory policy or individual company investments in security. For example, ransomware attacks occur over a broad range of industries [1]. Because of these structures, CPS design teams often do not regularly employ systems thinking, plan for system evolution, or long system life cycles. Thus, unwanted behaviors often go undetected in early testing or fielding but emerge in delayed feedback loops (often realized as zero-day exploits). Likewise, when vulnerabilities emerge, these systems are left open and vulnerable during their operational lifetime because there is little incentive to fix the problem(s).
- **Cybersecurity Mental Model:** The foundation on which iceberg events, patterns of behavior, and systems structure rest is the mental model for CPS design. The mental model represents the mindset and worldview of current CPS design teams. This mindset contributes to and builds all the layers above. For example, how software developers view security and privacy issues directly influences how they implement the SOI. Likewise, how the team manager views cybersecurity concerns will strongly influence the deliverable.

10.5 Reflection on Chapter 10 Material

In this chapter, we briefly presented several real-world applications of rigorous methods for systems thinking that are not yet commonly used by a broad range of functions in most organizations. The examples showed how the rigorous methods can enable us to apply the *systems principles* to improve our understanding and the odds of finding the best points of leverage on them.

Next, we move on to model-based systems thinking (MBST), an overarching rigorous method that integrates various modeling techniques that have proven very helpful in understanding real-world problems and finding their best points of leverage.

References

1. Beaman, C., et al. (2021). Ransomware: Recent advances, analysis, challenges and future research directions. *Computers & Security, 111*, 102490.
2. Bokhtier, G. M., Crawford, S., & Shahroudi, K. (2023). Integrated systems architectural modeling with architectural trade study of a UAV surface-less flight control systems for wildfire detection and communication utilizing MBSAP. In *Proceedings of the INCOSE International Symposium* (Vol. 33. pp. 229–250). https://doi.org/10.1002/iis2.13019
3. Bokhtier, G. M., et al. (2025). Systemic architectural optimization utilizing CFD analysis for flight controls with fan-in-wings (FIW) for wildfire UAVs/VTOLs. In *American Institute of Aeronautics and Astronautics (AIAA) SciTech Forum*, Orlando.
4. Checkoway, S., et al. (2011). Comprehensive experimental analyses of automotive attack surfaces. In *20th USENIX Security Symposium*, San Francisco. https://www.usenix.org/conference/usenix-security-11/comprehensive-experimental-analyses-automotive-attack-surfaces
5. Claroty Team82 (2021). *Biannual ICS Risk & Vulnerability Report: 2H 2021*. Technical Report.
6. Crawford, S. W. (2023). *MBSAP Application to UAV-based Wildfire Detection and Communication*. PhD Thesis. Colorado State University.
7. Lopez, N. (2021). *Copernicus: 2021 saw widespread wildfire devastation and new regional emission records broken*. Copernicus.
8. Speece, J., & Shahroudi, K. E. (2020). What we can learn from how service-oriented industries manage no-shows. *Journal of American College of Radiology, 17*, 412–413.
9. Sterman, J. D. (2000). *Business dynamics: Systems thinking and modeling for a complex world*. Irwin McGraw-Hill.
10. United States Department of Agriculture (2015). *The Rising Cost of Wildfire Operations: Effects on the Forest Service's Non-Fire Work*. https://www.fs.fed.us/sites/default/files/2015-Rising-Cost-Wildfire-Operations.pdf
11. United States Government Accountability Office (2021). *Weapon Systems Cyber-security: Guidance Would Help DOD Programs Better Communicate Requirements to Contractors*. Technical Report. Report to Congressional Committees.

Chapter 11
Model-Based Systems Thinking (MBST) Method

Abstract Model-based systems thinking (MBST) is a rigorous method (framework, languages, and tools) that integrates various modeling methods that have proven their value to augment our capability for understanding real-world problems but were not previously directly aligned with adding rigor to the application of systems principles of Chapter 5. Currently, the most useful modeling methods in MBST are system dynamics, stochastic, agent-based, AI/ML-based, and architectural modeling. MBST augments our limited capabilities to make real-world problems easier to understand. It is a force multiplier for applying the *systems principles* and *ways of the systems thinker* into our approach and behavior. MBST augments our approach regardless of whether the problem fits the STEM, non-STEM, sociotechnical, social science, or, in general, some transdisciplinary space that includes several core disciplines. The MBST method is an extended version of the system dynamic method by John Sterman (*Business Dynamics: Systems Thinking and Modeling for a Complex World.* Irwin McGraw-Hill, 2000). The MBST method was created to enhance the teaching and enforce the application of systems thinking principles (listed in Chapter 5) in the SYSE 532 Class taught at Colorado State University.

11.1 Introduction

It is essential to realize that the *systems principles* of Chapter 5 and the ways of the systems thinker of Chapter 7 are directly applicable to understanding real-world problems and systems. Also, some easier methods, like system mapping and causal loop diagrams, can be taught as early as middle school so people can start applying the philosophy, behavior, and approach to their challenges without learning more advanced methods like MBST.

Consider the complexity of volunteering with seven other authors to write a book whose chapters require tight collaboration between the authors and cannot be individually assigned. The authors can do the right things by applying the *SP*s and *SW*s. For example, they can all see themselves as part of the system, avoid win/lose mindsets, consider the longer-term impact of individual/team decisions,

K. Eftekhari Shahroudi et al., *Practical Systems Thinking*,
https://doi.org/10.1007/978-3-031-89394-0_11

and look for ways to make the integrated book much more than the sum of their contributions. They can attempt all this without adopting a more advanced framework, languages, and tools that make their understanding of the challenge and delivery easier. However, soon, the complexity of the interrelationship and flow of material and opinions will grow to overwhelm the authors, encouraging them to consider methods to augment or force multiply the team's ability to visualize the whole book (e.g., through mind mapping), to track and synchronize changes from eight remote authors (e.g., using Overleaf), to detect and fix punctuation, grammar, and voice style issues (e.g., using Grammarly), and to automate type setting and making the end product look good (e.g., using LaTex typesetting language), etc.

So, the difficulty of a real-world problem can be such that we would wish to have a force multiplier. Using MBST is a force multiplier method for systems thinking. It encourages the integrated use of various modeling techniques that were typically used in an isolated manner previously to understand, for example, just the architectural aspects (i.e., architectural modeling) or just the interrelational aspects (i.e., classical system dynamics) or just the human interactions (i.e., agent-based modeling), etc. This chapter explains the MBST framework that augments our transdisciplinary mental models via executable models using computers to support a better understanding, decision, and policy-making to impact real-world problems. MBST is also the antidote to practitioners of advanced systems requirements, design, analysis, verification, and validation tools who become good at using the tools but are not necessarily successful because their entire focus and proficiency are directed at cranking the tools rather than using the systems thinking principles to enhance their mental models.

MBST is defined as the *rigorous* application of *systems principles* to understand and solve *real-world* problems where the *rigor* comes from *discipline agnostic* modeling *methods*. MBST rigorously updates the *mental models* with high leverage *influence strategies* to improve emergent complex behavior of systems. We leverage these *influence strategies* to make changes to *purpose, structure, and patterns of behavior* of a system to enable better predictability, controllability, reliability, profitability, and other significant measures of our solutions and policies.

MBST enhances the systems thinker's appreciation of the interconnectedness of systems, willingness to embrace complexity, and commitment to continuous learning and adaptation. It has built-in features that intentionally limit traditional linear thinking. MBST is not just about using models; it rigorously enforces systems thinking through modeling artifacts that purposefully apply the systems principles listed in Chapter 5. It is an advanced capability that requires time and resources to aquire regardless of our classical backgrounds. For example, STEM disciplines do not have an easier barrier to entry than non-STEM disciplines. MBST is like a good GPS that makes the journey toward more sophisticated and effective systems thinking more efficient and better suited to the challenging interconnected information highways of the twenty-first century.

MBST promotes a more abstract way of thinking about real-world problems and systems that may influence them. It shifts the focus from text or picture-based descriptions to model-based representations, which are richer and better suited to handle complex systems. They say a picture is worth a thousand words. We say a model is worth a thousand pictures! Practitioners of MBST must adopt a mindset that is comfortable with complexity and uncertainty. This method involves recognizing patterns, understanding relationships, and anticipating the behavior of systems over time. It necessitates using numerous models, each providing different perspectives and insights. These can be descriptive, explaining systems function; predictive, forecasting future states; or prescriptive, suggesting actions to achieve desired outcomes.

The diversity of modeling methods used in MBST includes system dynamics, stochastic, agent-based, AI/ML-based, discrete event, and other modeling methods. Each modeling method serves a unique purpose and adds value to the system's understanding. For instance, system dynamics modeling is beneficial for understanding how reinforcing and balancing feedback loops with delays interact to drive a complex behavior that is not explainable by linear methods. Agent-based modeling excels at capturing the interactions of individual components within a system, and discrete event modeling is adept at modeling the flow of entities through a system.

Integrating these modeling methods allows for a more comprehensive understanding of systems. It finds the best leverage points where small changes can lead to significant impacts. Moreover, MBST encourages an iterative process of model building, testing, and refinement, essential for dealing with systems' evolving nature.

Mastering MBST augments the personal journey toward systems thinking described in Chapter 9 but can also impact the organizational journey toward a culture that fosters systems thinking. It requires collaboration across disciplines, as no single model or expertise can fully capture the complexity of real-world problems and the systems that may influence them. This interdisciplinary approach is crucial for the development of robust and resilient solutions.

MBST is not to be confused with Model-Based System Architecture Process (MBSAP) [1] or model-based systems engineering (MBSE) [6]. MBSAP and MBSE were initially mainly about architectural modeling applied to designing and engineering products and services that are more specialized, machinery, and process oriented than MBST.[1] MBST does not have a technical, product, or service-oriented bias and mainly uses *systems principles* and *ways of the systems thinker* to approach and understand real-world problems and systems that may influence them. So, MBST is more about understanding and learning real-world problems or finding the right problem than seeking solutions. Obviously, a primary benefit of better defining, understanding, and identifying real-world problems is finding solutions for them! However, the reader may have noticed that most of the material in this book is targeting the problem side more than solutions. This is not an accident. There are multiple reasons for this:

[1] Note all methods evolve and MBSAP and MBSE are evolving to include more than just architectural modeling.

First, our experience shows that highly detailed tools that are solution focused require so much energy and occupy so much of our cognitive bandwidth that they encourage us to quickly get onto and stay on the *solution treadmill* at the cost of spending enough time and effort on defining or understanding the problem. Here, running with quick solutions and cranking the tools that create attractive artifacts takes priority over applying the systems principles to define and understand problems so that we become efficient at creating artifacts or solutions that do not succeed. Pure solution focus tends to encourage compartmentalized thinking.

Second, we have found that the ability to define and understand the problem is a reliable indicator of how successful a project will be. Another way to say the same thing is that it is hard to deliver a solution, product, or service of value if we do not have the right *mental model* of the problem. This is why the MBST method explained in more detail in this chapter, has slow and fast learning loops, which allow us to adapt our mental models. The learning loops of MBST enforce the *learning principle* described in Section 5.9.

MBST does require a significant effort for personal mastery. A semester-long class is typically required as most of us do not already naturally possess the modeling skills and ways of thinking to guide the modeling effort. Giving in-depth training on MBST is beyond the scope of this textbook. We wrote this chapter to show what it is and how it *rigorously* enforces the *systems principles* of Chapter 5 and the *ways of the systems thinker* of Chapter 7.

11.2 The Guiding Principles of MBST

MBST enforces the systems principles listed in Chapter 5 so it encourages practitioners to view real-world problems as interacting systems and recognize their interactions, the interplay of their components, and all the influences that drive their complex behavior. As we follow the MBST framework, the visualization, modeling, and other artifacts force us to balance our efforts and focus between all the concerns of the systems principles of Chapter 5. This way, our actions, and behaviors are further aligned with the*ways of the systems thinker* listed in Chapter 7.

For example, MBST encourages looking beyond the immediate problem to see the larger patterns and structures that cause behavior. It encourages understanding how the system fits into the context of the larger environment. It also encourages understanding how humans and their mental models contribute to the behavior that we can observe.

The various rigorous methods of MBST serve as a shared language that can improve communication, collaboration, and alignment of mental models among stakeholders. For example:

– **System dynamics modeling** provides a way to explore system behaviors and test scenarios and their sensitivities in a virtual environment. The teams must understand the structure that drives the behavior of complex systems together with anticipation, prediction, or explanation of possible unwanted or undesirable behavior.

- **Architectural modeling** establishes the *3-systems* boundaries for various stakeholders, together with how the components are interconnected with stakeholder *purposes*, *roles*, and *responsibilities*.
- **Iterative incremental learning** adapts the various mental models and refines the definition of the problem as we cycle through the MBST framework. Each cycle updates the patterns of behavior, a high-level understanding of causal effects, and the integrated multidisciplinary model that shows how the various influences interplay to drive the emergent complex behavior.

11.3 The Model-Based Systems Thinking (MBST) Framework

The model-based systems thinking framework (Fig. 11.1) is an incremental and iterative process to understand real-world problems and the systems that may operate in, interact with, or influence them. It uses an integrated set of modeling methods to understand how systems' purpose, structure, and behavior influence each other.

MBST can be used to predict the emergent behavior of complex systems in science, engineering, politics, economics, and any complex phenomena that are difficult to understand or explain. It encourages a holistic view of complex phenomena instead of focusing solely on individual components. Model-based

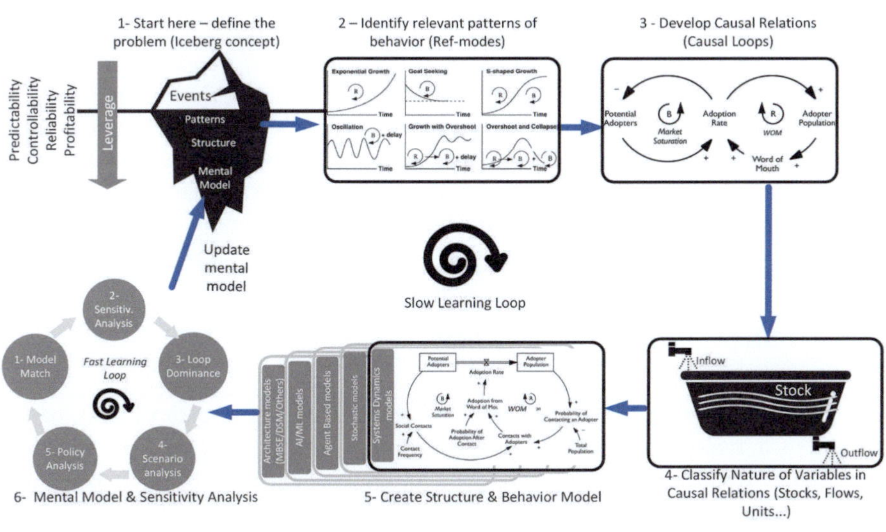

Fig. 11.1 Model-based systems thinking framework is an *iterative incremental process* that has *slow* and *fast* learning loops

systems thinking considers how elements within a system interact and influence each other. This holistic understanding helps to uncover hidden connections and dynamics.

MBST integrates several major modeling methods that have previously proven useful in understanding real-world problems and related systems.

Other significant or emerging modeling methods exist and can be integrated into the MBST framework. However, the selected modeling methods have been widely and reliably used to shed light on real-world complex problems for decades. They all enable a systems-oriented approach to understanding and solving complex problems consisting of:

The advantages of MBST are briefly:

- Identification of relevant *patterns of behavior* or useful information from data
- Understanding the causal *structure* that drives the *patterns of behavior*
- Understanding how the *purpose architecture* and *behavior* of systems influence each other
- Locating best points of leverage to influence *purpose architecture* and *behavior*
- Predicting potential future scenarios and unwanted long-term impact of systems
- Updating and aligning mental models of individuals, teams, and organizations
- Not biased toward a particular type of real-world problem or system: i.e., discipline agnostic

Our teaching experience shows that regardless of our educational background or discipline, most students do not naturally possess the competencies to complete the MBST framework's various steps. So, training is needed to utilize the MBST framework. This book is a helpful high-level guide or map to inform our personal and organizational journeys toward systems thinking. However, it is introductory and not a replacement for proper training on all the required competencies for MBST.

The six steps of the *slow learning loop* of the MBST framework of Fig. 11.1 and the *fast learning loop* of the last step are described in the following subsections. Each step is valuable for approaching problems, even if we are not trying to adhere to MBST framework! However, doing the learning iterations rigorously enforces all the *systems principles* and strongly enhances our ability to understand real-world problems and the systems that may influence them.

11.3.1 Step 1: Define the Problem

The first step is to define or identify a problem. This is any data or information related to a real-world problem, obstacle, issue, or behavior we are trying to understand. To define a problem, we need to define or question the *why, what, and how* behind any of the layers of the iceberg shown in Fig. 11.1. Let us say that we are concerned with the problem of homelessness in big cities. This is an example of a way to specify the problem:

– **Event**: Why is the visible homeless population growing rapidly in urban areas like New York City or San Fransisco despite significant growth in expenditure to contain it?
– **Patterns of Behavior**: How do the patterns of behavior of homeless counts compare or relate between large cities and smaller cities with populations less than 500,000?
– **Structure**: How are the known patterns of behavior influenced by current policies and solutions?
– **Mental Models**: What metrics do cities use to reward or pay organizations that are delivering products or services to contain or improve homelessness?

Of course, our current mental models of what causes homelessness will color or shape the questions we ask, but very often, they will likely be wrong, or they may require a significant adjustment. If the prevalent mental models of city planners and policymakers were perfect, then their investment would interfere with the problem at the best leverage points, and we would see a sustained drop in the homelessness phenomena over the long term. So, it is not required that our problem definition questions are perfect or the best because the whole point of the *slow and fast learning loops* of the MBST framework of Fig. 11.1 is to update our mental models of the problem so we can keep refining the problem definition until we have mental model that has a significant, sustained long-term impact at a reasonable cost. Now, the *event* layer of the problem iceberg is the easiest to observe or measure because that is the visible part above the water. The deeper layers are harder to understand since they are not directly visible or measurable. The MBST framework aims to understand the deeper layers and how they are interconnected and influence each other. The better we understand how the problem iceberg layers influence each other as we go through the iterative and incremental *slow and fast learning loops*, the more predictable, controllable, reliable, and profitable (or cost-effective in this case) our understanding and therefore our solutions and policies will be. How do we know this is true? Well, the *learning principle* states:
Learning principle: (common) *mental models* influencing human *decisions* about *architectures* are *learnable* and *adaptable*.

Historically, systems thinking textbooks by Meadows [7]and Senge [8, 9] and even some sections of the Business Dynamics by Sterman [10] appear to suggest a linear chain of cause and effect from *mental models* to *purpose* to *structure* to *patterns of behavior* to *events* that is the visible part of the problem. But as the right side of Fig. 5.15 in Chapter 5 showed, learning helps us go deeper or in the other direction. So, the modern way to think about the problem iceberg is no longer that the best (or most practical) leverage point to focus our investment or interference is always at the deepest layer. In this book, we contend that the layers of the problem iceberg are interrelated and influence each other: upward, downward, and in any other possible direction where multiple layers can directly interact. The better we understand the layers and their influence on each other, the better our chances of finding the best and most practical point of leverage that may be located anywhere on the problem iceberg.

However, the problem iceberg is still a helpful visualization tool that ensures we consider all the layers of the problem rather than only focusing on a particular aspect.

11.3.2 Step 2: Identify Relevant Patterns of Behavior

The second step involves identifying the relevant behavior patterns using visual or more involved statistical methods. This is typically achieved by plotting the *event*, the measurable or visible part of the problem over time, and detecting, recognizing, or extracting typical patterns of behavior, six of which are shown in Step 2 of Fig. 11.1. We sometimes refer to the patterns we often see as *behavioral archetypes* because they are commonly observed in many problems. An actual response will typically exhibit several of these *behavioral archetypes* at different periods or a combination of several simultaneously. The *behavioral archetypes* are exponential growth (or nose dive), goal-seeking behavior, S-shaped growth, oscillation, growth with overshoot, and overshoot with collapse, etc. Identifying *behavioral archetypes* in the time plots is essential because they will make creating the *causal relation* in the next step of the MBST framework easier. Sterman [10] teaches us that every *behavioral archetype* has a simple and known *structural archetype*. For example, overshoot and collapse occur in a closed champagne bottle where yeast cells consume sugar and grow in numbers until their population rises above the level that is sustainable by the available sugar resource. As the population continues to consume and deplete the sugar level in the bottle, it becomes harder for the yeast to survive, so they start dying. This overshoot and collapseare a known *behavioral archetype* for which there is a known *structural archetype* with one reinforcing and two balancing loops. The other well-known archetypes are:

- **Exponential growth** or nose dive that looks like a mirror image (or inverse) of the exponential growth can be minimally represented by a reinforcing or R loop.
- **Goal-seeking behavior** (e.g., as observed in policy resistance situations) can be minimally represented by a balancing or B loop with a gap variable inside the loop receiving a command or input from a goal variable.
- **S-shaped growth** (or an inverted S-shape of loss) can be minimally represented by a goal-seeking behavior with an additional reinforcing loop: B loop with a gap variable inside the loop, with a goal variable from outside the loop augmented with an R loop.
- **S-shaped growth with oscillation** can be minimally represented by adding a significant delay inside the B loop of **S-shaped growth**

11.3.3 Step 3: Develop Causal Relationships

The purpose of Step 3 is to identify the key variables that may have a relevant causal relationship to the patterns of behavior identified in Step 2, in addition to how they influence each other. The resulting artifact is called a causal loop diagram (CLD). Chapter 7 and then Chapter 5 of the Sterman textbook [10] are good starting points to learn about CLDs. CLDs are not only concerned with linear causal chains, as seen in a fishbone diagram like that shown in Fig. 1.2. A fishbone diagram intends to capture independent causes of some outcome, but there is no encouragement to understand if the causes may have an interrelationship or interinfluence. That is why we do not see any feedback from skills shortage to any causes in Fig. 1.2. By contrast, CLDs encourage us to understand all the *relevant* circular or feedback causal structures like reinforcing R loops and balancing B loops. Comparing Figs. 8.8 to 1.2 shows how systems thinking differs from linear compartmentalized thinking regarding cause and effect. By *relevant*, we mean causal structures (i.e., CLDs) that can explain the patterns of behavior identified in Step 2.

With some training, utilization of known archetypes, and referring to existing examples like those given by Sterman [10], we can create a good first version of a causal loop diagram that can potentially explain the patterns of behavior identified in Step 2. CLDs are essential because they are a first attempt at understanding the behavioral aspects of the real-world problem structurally. It is also an essential step that enables Steps 4 and 5 to create rigorous executable or running models on a computer.

11.3.4 Step 4: Classify Nature of Variables in Causal Relations

We know examples of research and development projects where highly regarded and well-known engineers wasted millions of dollars trying to correlate variables because they did not have the urge or training to understand the nature of variables. For example, let's imagine that there is value in understanding the location and amplitude of peak pressure in a tank used in a large chemical reactor process, where the value comes from protecting the tank from damage, in addition to the efficiency and quality of the reaction process the tank pressure is used for. Next, let's imagine that this pressure sensor is super expensive, costing several times more than the tank it is trying to protect. Now, let's imagine that there is another cheap and reliable sensor that does not directly measure pressure. Instead, it measures the net rate of material flow into the tank. So, what it measures is related to the pressure in the tank: if the net flow rate into the tank is positive, then the pressure goes up. The pressure goes down if the net flow rate into the tank is negative. Suppose we do not have the intuition or framework that distinguishes between stocks and flows. In that case, we may correlate the location and amplitude of the peak net flow rate with the location and amplitude of the peak pressure. Unfortunately, there is a structural reason why this approach will never work, regardless of how much testing and money we pour into this activity to enable the cheaper net flow rate sensor to achieve the same as the pressure sensor. When systems thinking becomes intuitive to us after training, we would naturally know the reason: the location and amplitude of the peak pressure do not directly relate to the location and amplitude of the net flow rate. The location and amplitude of the peak pressure match the zero crossings and the area under the curve of the net flow rate. Even highly paid and respected engineers can make the mistake of going after futile endeavors because they have not learned to step back and classify the nature of variables!

Chapter 7 discusses the difference between stocks (variables that represent accumulation like a bathtub) and flows (variables that represent a rate of change of stocks). First, we need training on stocks and flows to understand their relationship visually, graphically, and intuitively. Then, we can ask and answer the question of whether variables in a CLD are stocks or flows or other categories such as discrete, real, integer, complex, lookup, exogenous, time based, frequency based, stochastic, disturbance, etc.

This step also helps bypass the mathematical hurdle of specifying the problem in ordinary differential equations, differential algebraic equations, or other mathematical or numerical forms beyond most people. We found that even the people comfortable with using math software do not intuitively understand the behaviors they compute, so they are more likely to accept the erroneous results of a simulation. Most backgrounds feel like fish out of water when understanding and specifying problems directly in mathematical equations. Classifying the nature of variables helps us bypass this hurdle because software like Stella and Vensim hide the mathematical and numerical processes under the hood as long as we can classify the nature of variables.

The fourth step is the beginning of building a dynamic model for simulating the behavior patterns resulting from a known structure with stocks, flows, interacting reinforcing, and balancing loops where all variables have been appropriately classified. We note that Steps 1 to 4 are common regardless of which combination of rigorous modeling methods we will employ at the next step.

11.3.5 Step 5: Create Structure and Behavior Models

Using a systems thinking lens to view real-world problems as systems or systems that may influence them, then the purpose of the fifth step is to rigorously understand and discover the relationship between all the significant concerns of systems emphasized in italic text in the statement of *systems principles* in Chapter 5.

This step involves using various rigorous modeling methods, whether isolated or integrated. The following modeling methods have all proven their value for their ability to shed light on different aspects of real-world problems to make them easier to understand:

- **System Dynamics Modeling** (Sterman [10]): Focused primarily on the interaction of all the features and concerns of the *holism*, *architecture*, *emergence*, and *human* principles
- **Architectural Modeling** (Borky et. Al [1]: Focused primarily on the interaction of all the features and concerns of the *architecture*, *holism*, *3-systems*, and *boundaries* principles
- **Stochastic Modeling** (Franzke et. Al [4]): Generic focus on all systems principles, plus the ability to create pieces that fit into or support the other modeling methods
- **Agent-Based Modeling** (Chen et. Al [2]): Focused the interaction of all the features and concerns of the *HumanSp* and *EmergenceSp* where complexity is due to single, team, or organizations of humans
- **AI/ML Modeling** (Gorriz et. Al [5]): Generic focus on all systems principles, plus the ability to create pieces that fit into or support the other modeling methods

There is excellent reference material for the above rigorous modeling in the reference section of this introductory textbook, plus classes we can take to learn the above methods. Chapter 12 shows some example artifacts created by these modeling methods and discusses how creating these artifacts enables or enforces the principles of the system. A word of warning, it is quite possible, for example, to create an architectural model, only to get sucked into playing with the tool versus applying systems thinking to understand the problem better. The effectiveness of these methods depends on how rigorously they are used to aid systems thinking. We have seen many cases of tools becoming the end goal. Becoming proficient in the above modeling tools can occur at the expense of knowing and applying the systems principles. So, it is possible to become an expert-level user of these modeling methods where the user's bandwidth is fully occupied with cranking the tools versus applying the systems principles to understand real-world problems. We believe that these modeling methods are used to enforce the systems principles to augment our human limitations in understanding real-world problems. However, we know these tools can also become their primary focus, reinforcing counterproductive linear compartmentalized thinking. This is why we recommend teaching these modeling methods as systems thinking enablers rather than stand-alone machinery.

11.3.6 Step 6: Mental Models and Sensitivity Analysis

Once we have a running model that we can simulate on a computer, the first activity is to match the patterns of behavior (also called reference data or reference modes) that we identified in Step 2. This means that we update, adjust, or refine the structure of the model until its output matches the reference modes. Without this step, we will have no confidence that our models have the correct or helpful structure to illuminate the real-world problem we specified in Step 1. Once the model is matched to the reference mode data, it enables *slow and fast learning loops* described in the next section. Both loops intend to update the problem definition of Step 1, including our mental models. The *fast learning loop* attempts to squeeze and capture as much value as possible from the matched model relatively quickly via sensitivity analysis, scenario analysis, and policy/control analysis. The *slow learning loop* is when all the value is squeezed out of the matched model, and to gain even more understanding, we now have to repeat the main steps of the MBST framework. This will be discussed in more detail in the next section.

11.3.7 Incremental Iterative Update of Mental Models: Slow and Fast Learning Loops

Figure 11.1 shows that we update our mental model at the end of every iteration of the *slow learning loop*. This allows us to refine, redefine, or reframe the problem we are trying to understand. This can, for example, require new data and the identification of new patterns of behavior, new causal relations, model development, etc. We call this a slow loop because updating our mental model causes every step of the MBST framework to require a significant update potentially. All the steps in Fig. 11.1 require a dedicated effort, mainly if the real-world problem chosen is very complex. For example, we know from previous teaching experience that the skills required for system dynamic modeling do not come naturally to most people: in-depth semester-long training is typically required, but depending on the complexity of the problem, even an expert-level system dynamicist will put in a significant amount of time and effort to develop a model (Step 5) that matches the reference patterns of behavior identified in Step 2. Going from Step 1 through the MBST framework back to Step 1 is considered a *slow learning loop*.

Once a running model is developed that matches the reference patterns of behavior of Step 2 in Step 6, then a *fast learning loop* is typically possible where one can relatively quickly squeeze a lot of value and understanding by exercising the model:

- Sensitivity Analysis: Finds the sensitivity of the complex behavior to parametric variations
- Loop/Structural Dominance: Finds which parts of the structure dominate what part of the complex behavior
- Scenario Analysis: Prediction/extrapolation of potential future behaviors past the known recorded behavior in Step 2
- Policy Analysis: Adding policy/control structures to the model to test their effectiveness in the near and long term

The *fast learning loop* is typically more fun as a lot of insight is gained relatively quickly, such as finding the best point of leverage to interfere with the system to get a desired behavior or predicting undesirable longer-term side effects of implementing a new policy, etc. Furthermore, the *fast learning loop* does not require in-depth training on the rigorous previous steps, so it is, for example, possible for an expert to accomplish a *slow learning loop* and then hand over what is called a *management flight simulator* to a casual user to play with and learn from it.

11.4 Example of Using MBST to Understand Max Transport Ridership

The MAX Transport shown in Fig. 11.2 started service in May 2014 as the first rapid transit system in Colorado. Three CSU students who were in contact with the city planners followed the MBST framework of Fig. 11.1 although we did not have a good name like MBST for the framework at the time. They built a system dynamics model of MAX Transport ridership to understand the factors that impacted ridership. This model is shown in Fig. 11.3. Their model helped shed light on the following:

- The rapid rise, fall, and plateauing of ridership after introduction experienced in a similar transit system introduced in Minneapolis earlier
- The complexity of ridership growth
- The complexity of attractiveness
- The influence of public and private transport on each other (Note: for brevity, Fig. 11.3 does not show the private car ridership part of the model)

Most importantly, their approach helped them understand the impact of the following on MAX ridership: advertising, word of mouth, quality, capacity planning, the opening of a new east-west line, where to invest money to escape the ridership plateau, etc.

Initially, MAX ridership grew rapidly for the first 8–12 months after its introduction into service. But this growth did not sustain and it quickly plateaued to 100,000. So, the first question was: what if we increase the number of buses? Figure 11.4

Fig. 11.2 MAX Transport, which started service in May 2014, was studied by three students in Dec 2014 at CSU to understand ridership behavior before investment to expand ridership

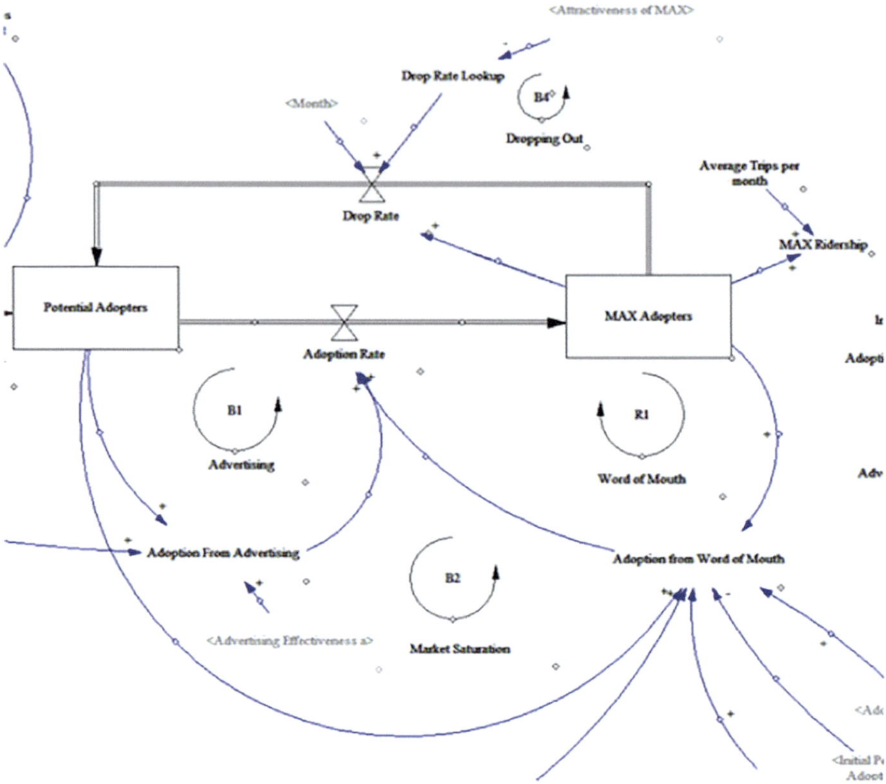

Fig. 11.3 Adoption loops part of a system dynamic model of the MAX Transport

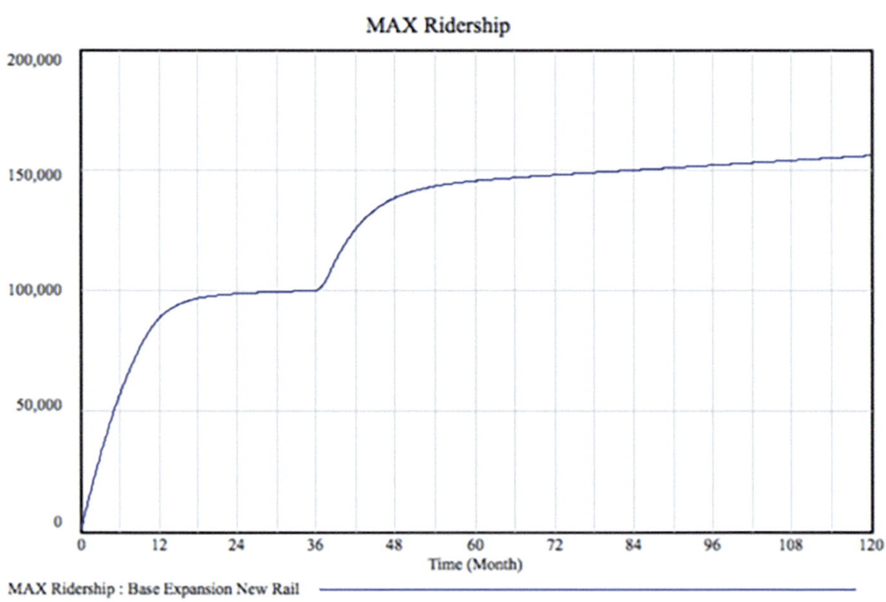

Fig. 11.4 Predicted response of doubling investment into more busses showed that ridership would not double

shows their model's prediction for doubling bus investment. The model predicted that doubling the investment alone was not the best leverage point because Max ridership only increased by 50% and then transitioned to a slow growth that was more a function of the slow predicted growth of the population of Fort Collins versus any other effect above. After this point, we were not privy to the detailed executive decisions resulting from their improved mental models, but we know that their policies led to MAX ridership reaching 450,000 by 2019 before the COVID-19 pandemic based on publicly available information.

11.5 Differentiating MBST Versus Other Model-Based Methods

We started using the term MBST in the SYSE 532 class at the Colorado State University because the other system dynamics (SD), Model-Based System Architecture Process (MBSAP), model-based systems engineering, and many other methods that start with "model based", while related, did not quite capture the intent of MBST to align and integrate various modeling methods for systems thinking: here, the tools often became the goal in themselves, absorbing all the bandwidth of practitioners where they were no longer explicitly enabling systems thinking. We added this section to help clarify MBST while fully acknowledging the value and benefit of other model-based methods.

Systems thinking (ST) considers real-world problems as systems by encouraging us to consider systems principles features such as elements, interconnects, purpose, structure, behavior, emergence, etc. Model-based systems thinking (MBST) integrates multiple modeling methods to augment our capabilities in this endeavor. However, the integration forces the modeling methods to explicitly function within the meaning and intent of the systems principles.

11.5.1 MBST Versus SD

MBST is a particular interpretation and expansion of system dynamics (SD) taught by John Sterman [10]. MBST gives an explicit process that includes fast and slow learning loops. It also expands the scope of SD to integrate other systems-oriented modeling methods, such as architecture modeling, agent-based modeling, AI/ML modeling, etc. The advantage of integrating the other modeling methods into MBST is that it forces the other methods to be applied in a systems-oriented way. By integrating the other modeling methods into MBST, we constrain them to apply systems principles. For example, when we integrate AI/ML modeling into the MBST framework, it can no longer be used to make linear compartmentalized thinking even more efficient. So, the essential difference between MBST and SD is that MBST is an explicit umbrella for integrating while at the same time constraining various other modeling methods to apply the systems principles to understand better real-world problems and the systems that interact with them.

System dynamics is one of the modeling methods that MBST integrates. It remains a niche expertise and has not penetrated the industry far beyond boutique management consulting firms. Creating and utilizing a validated and verified SD model that continues to deliver value over time continues to be challenging. The MBST framework utilizes other modeling methods to make the rigorous application of systems thinking easier. In Section 11.6, we explain how technology trends will help MBST penetrate the industry more significantly than SD did.

11.5.2 MBST Versus MBSAP/MBSE

The essential difference between MBST and the Model-Based System Architecture Process (MBSAP) and its cousin, model-based systems engineering (MBSE), is that MBST today is more about understanding real-world problems. On the other hand, MBSAP and MBSE help us create products and services that interact, operate in, or interfere with them. Understanding and creating are different activities! However, one without overlapping the other will not work. For example, buying the best MBSAP/MBSE tools does not automatically deliver successful products and services if cranking the tools becomes the central focus versus applying the systems principles.

Now, we know that MBST, MBSAP, and MBSE evolve because the *evolution principle* states:

Evolution principle: systems have *life cycles* and they *evolve*.

Hence, in the future, the purposes of MBST, MBSAP, and MBSE may expand to merge, making it difficult to compare and contrast them. However, today, they have clearly differentiated emphasis.

11.6 Impact of Technology Trends on MBST

All real-world problems and the systems that operate in them or interact with them in some way, even the relatively simple ones with very predictable or logical patterns of behaviors, can suddenly become complex in the presence of change, anomalies, uncertainties, or disturbances. While we are just beginning to understand the human brain, the complexity of problems we face quickly appears to exceed our unaided cognitive individual, group, and organizational capabilities. For example, even the best fighter pilots can only keep track of a handful of moving targets or different types of adversity if unaided by systems that make handling their missions easier. So, there is a need for external frameworks, languages, and tools that help augment our limited capabilities to make it easier to understand and deal with real-world problems.

The MBST method helps because it rigorously enforces the systems principles and the rigorous modeling methods enhance our limited human capabilities to deal with complexity. However, the personal and organizational journey toward systems thinking culture is not easy, and we are actively researching ways to further automate and enhance the speed of the two learning loops in Fig. 11.1. We are seeing progress that suggests artificial intelligence will reduce this hurdle by augmenting the human systems thinker with the following tasks:

– **Collecting** reference mode data that is relevant to the definition of the problem in Step 1
– **Identifying** the patterns of behavior in the reference mode
– **Creating** an initial CLD that has a structure to explain the above patterns of behavior
– **Adding** stocks and flows to create a first executable model
– **Filling** cause-and-effect equations, lookup tables, and correlations
– **Verifying** and **validating** the model to match the reference mode data

Once we have a running verified and validated model, AI can help us more quickly identify which parts of the structure in the rigorous models of Step 5 dominate which patterns of behavior in the reference mode data. This will be valuable, as it will help find the best leverage points to modify the behavior.

In summary, AI/ML technology will soon boost the MBST method by automating the initial iteration of the slow and fast learning loops, given a problem definition (Step 1 of Fig. 11.1 framework). After the initial iteration, the human systems thinker steers the subsequent iterations by requesting high-level or detailed changes to one or more steps while continuing to understand the complexities better and identifying practical leverage points on the real-world problem and the systems that intend to improve it in some way.

Note that the above vision of the near future maintains all the artifacts in a human-comprehensible and steerable form. We may challenge this vision and ask this reasonable question: why not use AI to bypass the human systems thinker entirely and come up with problem definitions and solutions that are not understandable or steerable by humans at any level of abstraction? But in this case, how would the human systems thinker, presumably responsible for the quality of the vast set of AI-created problem/solution sets, independently validate and verify that they are practical, cost-effective, and sustainable and cannot cause more severe problems in the future? So, our answer is that we need MBST to remain human-comprehensible and steerable because the *human principle* requires it:

Human principle: *humans* impact and are impacted by the (wanted/unwanted) *emergence* of *all* reachable systems.

11.7 Reflection on Chapter 11 Material

Here, we introduced the model-based systems thinking (MBST) method that rigorously enforces *SP*s and *SW*s. The MBST method is an extended version of the system dynamics [3, 10] method. It integrates several modeling methods in addition to system dynamics to increase our ability to see real-world problems as systems. MBST improves our understanding and helps us find the best leverage points in our mental models, structure, and behavior of systems. This then translates into increased predictability, controllability, reliability, profitability, and other significant figures of the merits of our solutions and policies. The next chapter takes a deeper dive into the use of some of the rigorous modeling methods that are currently under the MBST umbrella.

References

1. Borky, J., & Bradley, T. (2019). *Effective Model-Based Systems Engineering*. Springer.
2. Chen, J. J., Zheng, B., & Tan, L. (2013). Agent-based model with asymmetric trading and herding for complex financial systems. *PLoS ONE, 8*(11). https://doi.org/10.1371/journal.pone.0079531
3. Forrester J. (1992). *From the Ranch to System Dynamics: An Autobiography*, A. G. Bedeian (Ed.). Management Laureates (Vol. 1). JAI Press.
4. Franzke, C. L. E., et al. (2022). Stochastic methods and complexity science in climate research and modeling. *Frontiers in Physics, 10*. https://doi.org/10.3389/fphy.2022.931596
5. Górriz, J. M., et al. (2023). Computational approaches to explainable artificial intelligence: Advances in theory, applications and trends. *Information Fusion, 100*, 101945. https://doi.org/10.1016/j.inffus.2023.101945
6. Long, D. A., & Scott, Z. (2011). *A primer for model-based systems engineering* (2nd ed.) . Vitech Co. ISBN: 978-1-105-58810-5
7. Meadows, D. H. (2008). *Thinking in systems: A primer*. White River Junction: Chelsea Green.
8. Senge, P. (2006). *The fifth discipline: The art and practice of the learning organization*. Random House Books.
9. Senge, P. M. (1994). *The fifth discipline fieldbook: Strategies and tools for building a learning organization*. Crown Currency. ISBN: 978-03-85472-56-2.
10. Sterman, J. D. (2000). *Business dynamics: Systems thinking and modeling for a complex world*. Irwin McGraw-Hill.

Chapter 12
Models for MBST Framework

Abstract This chapter explores the crucial role of modeling methods in MBST for understanding complex real-world problems viewed as systems. Modeling improves our *mental nodels* by capturing how the *purpose*, *structure*, and *behavior* of *systems* are interconnected and influence each other. The modeling methods discussed in more detail are architectural(MBSE and DSM), system dynamics, agent based, stochastic, optimization, and AI/ML. Here, we emphasize the importance of integrating systems thinking principles into these models to understand difficult to predict issues and improve our decision-making processes. We include practical examples and case studies, such as wildfire UAV fleet deployment and urban water and energy management, to give the reader a feel of how the various modeling methods help. We close the chapter and the book by discussing the potential benefits and challenges of using AI/ML (circa fall 2024) in systems thinking, advocating for its use as a supplemental tool to enhance the understanding and analysis of complex real-world problems viewed as systems.

This chapter takes a deeper dive into the various rigorous models used by systems thinkers for model-based systems thinking. The model-based systems thinking (MBST) method was introduced in Chapter 11 and shown in Fig. 11.1. Figure 12.1 focused on the modeling methods currently included in MBST. It shows different core types of models in the MBST framework used to explore the purpose, structural, and behavioral perspectives of a system. Modeling supports the rigorous application of systems thinking principles discussed in Chapter 5. Entire textbooks and a vast amount of literature exist for each modeling method that offers distinct advantages. Here, we focus on explaining how modeling helps systems thinking by enforcing the application of systems principles.

This chapter begins with **architectural modeling** including MBSE and DSM efforts. Architectural models are very useful for understanding how the *3-systems* are interconnected internally and with each other. We then move to how **system dynamics modeling** helps complex problem-solving. This is a very rich method that made it possible to create the MBST framework of Fig. 11.1 to integrate all the other modeling methods discussed in this chapter. System dynamics offers the utility for

K. Eftekhari Shahroudi et al., *Practical Systems Thinking*,
https://doi.org/10.1007/978-3-031-89394-0_12

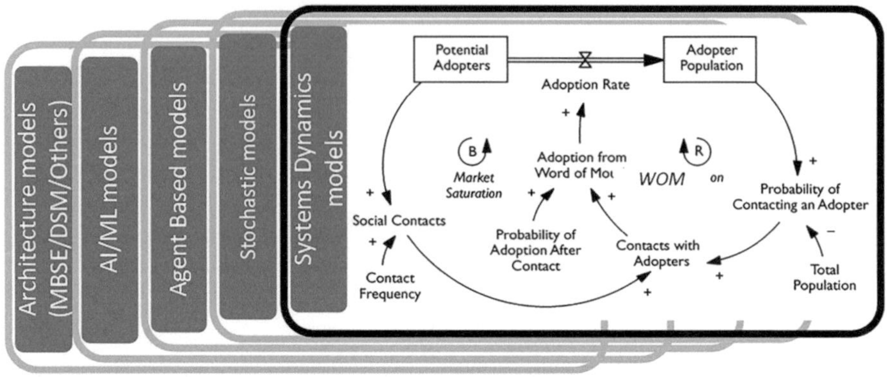

Fig. 12.1 Modeling methods integrated the MBST framework

understanding dynamic patterns of behavior over time with feedback, delays, and system responses. **Agent-based modeling (ABM)** simulates autonomous agents' interactions within a system, showing complex, emergent behaviors from simple individual rules, while **stochastic modeling** account for randomness in systems, using probabilities to represent variability and unpredictability in outcomes. **Optimization modeling** enhances resources, processes, or system configurations in various scenarios. **Artificial intelligence/machine learning modeling** provides an automated way to learn the complexities of a nonlinear system. They can be used to understand the interactions of multiple, interconnected parameters of a system that affect system behavior. These modeling methods provide a rigorous approach to understanding system complexities and are invaluable for implementing systems thinking.

12.1 Designing Architecture Models with Systems Thinking

Architecture models are one of the most effective systems thinking tools, as they help define the structure and interrelationships of systems with their environment and within their own internal subsystems and components. In the following sections, we will explore two of the most commonly used models for system architecture and systems thinking: model-based systems engineering (MBSE) and design structured matrix (DSM).

Figure 12.2 shows how different types of architectural models fit together and are enhanced by systems thinking principles.

Architecture principle: the *purpose*, *structure*, and *behavior* of systems *influence* each other.

The architecture models that we develop using MBSE, DSM, or any other modeling mechanism need to adhere to the architecture principle and other systems

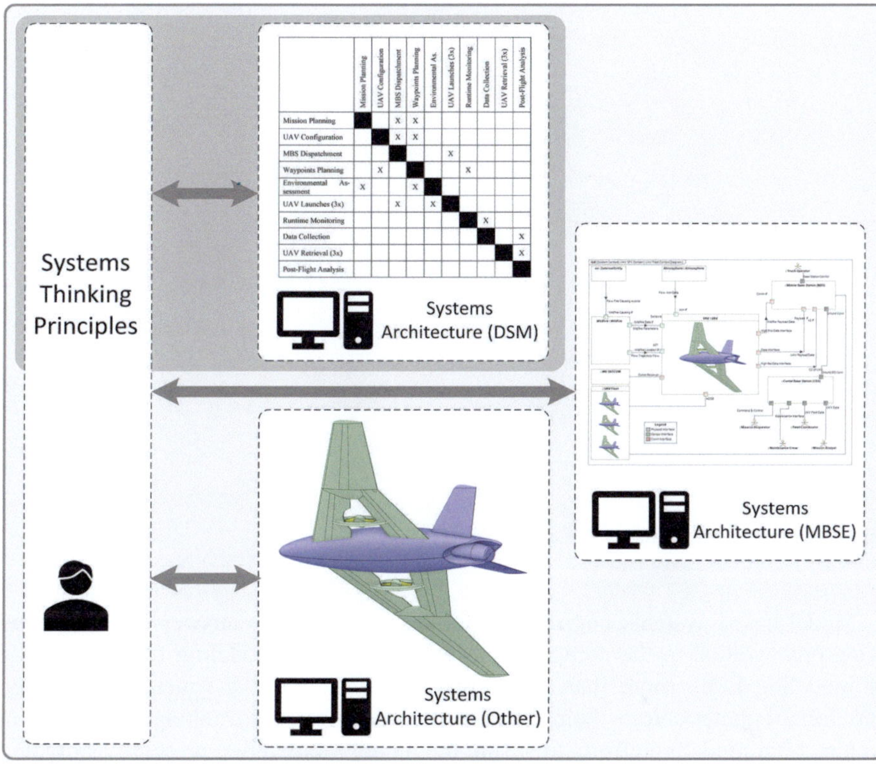

Fig. 12.2 Architecture models and systems thinking principles

principles. Effective architectural models built in MBSE or DSM consider the system's purpose, evaluate the system's structure, and characterize the system's behavior overall.

12.1.1 MBSE as Architecture Modeling Tool

MBSE is a critical development in advancing systems engineering and system design. Its implementation and proliferation have been widely adopted across industry and government as a best practice for complex system design and development. Systems modeling and architecture have evolved, with formalized languages bringing standardization to this effort. Specifically, MBSE utilizes the SysML modeling language, which is built on the unified modeling language (UML) foundations.

MBSE helps you rigorously navigate development, standardization, and communication through a single unified system representation, termed a "model." This model presents different viewpoints of a system and defines a system's architecture (Fig. 12.3).

Fig. 12.3 System interconnectedness

Model-based systems engineering is best used to help answer questions and bring organization to the complexity of system design. Modeling for the purpose of modeling is no more than generating pretty pictures. A structured language and formal methodology help address this, but systems thinking is critical to architectural modeling efforts to ensure the model accurately represents reality and provides utility to the design effort.

This section will focus specifically on MBSE architectural models, specifically system architecture's utility for systems thinking.

System architecture is defined as an abstract representation of a complex system/entity specified in terms of:

- *Structure*—how is the system organized; where do important interfaces occur?
- *Behavior*—how do individual pieces and part interact, and how do they collaborate to produce the overall behaviors that implement the functionality required?
- *Rules*—what are the conventions (e.g., standards and requirements) that must be adhered to in order to achieve a useful and satisfactory solution?

System architecture and MBSE helps with:

- Understanding and capturing the system purpose and requirements
- Ensuring that the structure, behavior, and configuration of the solution deliver the needs established in the system requirements
- Applying systems thinking principles in a rigorous method when developing a solution to a complex problem
- Adding rigor, machinery, language, tools, and process (framework)

MBSE provides inherent rigor, consistency, efficiency, and fidelity to systems thinking principles.

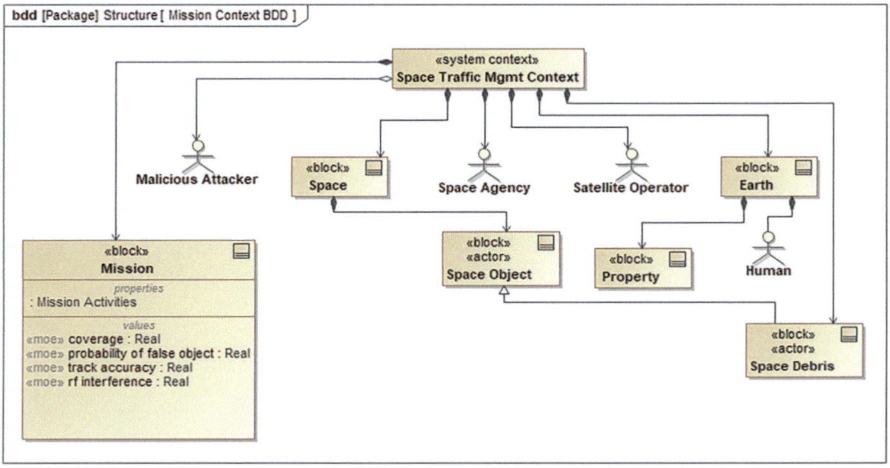

Fig. 12.4 Context diagram—external

MBSE with Systems Thinking

MBSE models, particularly those developed in any tools that have advanced analysis features (i.e., cameo, rhapsody, enterprise architect, Polarion Capella, MATLAB/Simulink, etc.), provide a robust framework for defining and analyzing system architecture.

Common System Architectural Models:

- Context diagrams present a structural view of a system—looking inward at subsystems and components. They also can model outward: illustrating the 3-systems principle of every system operating in its environment with other systems. Figures 12.4, and 12.5 are examples of context diagrams.
- Use case diagrams capture the key functionality of a system. They model the system actors, both operators and other systems interacting directly with the key functionality. Figure 12.6 is an example.
- Activity diagrams are functional models illustrating a sequence of actions that occur to fulfill intended functionality. These are very useful for a high-level overview of functions needed for a given system use case. They assist in systems thinking by modeling condition-based relationships and the flow of activities within a system's intended behavior. Figure 12.7 is an example of an activity diagram.

Example Architectural Models This section presents samples of architectural models representing a notional space traffic management system. This system would be responsible for tracking, identifying, and predicting space objects in order to provide efficient space utilization and prevent collisions.

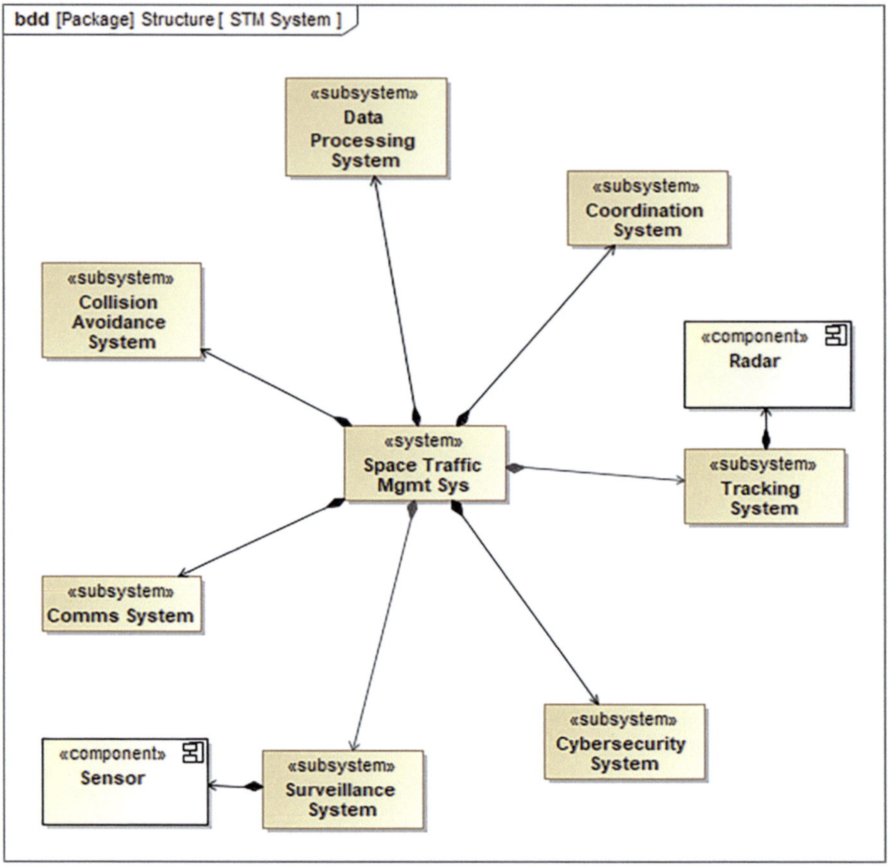

Fig. 12.5 System structure—internal

A *context diagram* highlights the system environment and other systems it interacts with in that environment. It is an excellent and simple way to visualize the 3-systems principle. Figure 12.4 shows the external systems and actors that the system design must account for. It models the relevant environmental considerations and various user types that will interact with the system as well. In Fig. 12.4, you can see the system of interest is the space traffic management system. It operates within the 3-systems principle with space as its environment and earth as an enabling system considering the gravitational pull and laws of physics that apply to the orbital mechanics. This also illustrates the human principle in capturing human actors that will impact or be impacted by the system.

An internal view of a context diagram, a system structural block definition diagram, as shown in Fig. 12.5 illustrates the holism principle. Here, the space traffic management system as a system of interest is shown as a composition of various elements and interconnects that interact to achieve a given purpose.

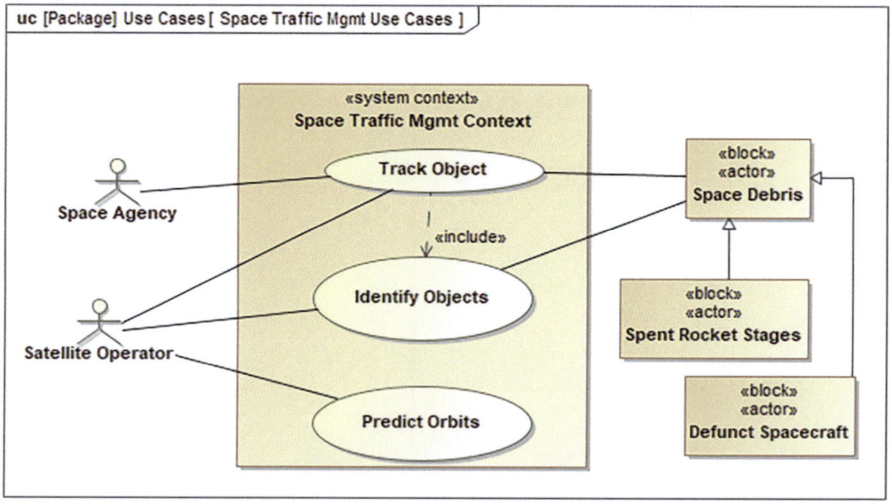

Fig. 12.6 Use case diagram

The use case diagram for the space traffic management system, shown in Fig. 12.6, illustrates the actors and primary functions the system is responsible for executing. It also shows relevant external actors that interact as a part of the primary functions. Here, the architecture principle is shown, the space traffic management system, having functional interconnections with space debris and also with the two types of human actors, the operator and the space agency users. It is more than just structural relationships but also behavior relationships to achieve the key functions of track object, identity objects, and predict orbits.

Activity diagrams show the flow of activities required to execute a system use case, such as tracking objects, shown in Fig. 12.6. They model the conditional logic required to execute the given activity and allocate actions to a given actor. In this case, the actions are separated between the space traffic management system itself and the user. Activity diagrams provide a functional viewpoint to utilize the emergence principle in considering the complexity of interactions impact on our system. It shows the flow of events and conditional logic of actions across various system elements and their interconnectedness to achieve a desired purpose.

The examples above are presented to demonstrate the modeling of architectural modeling as a rigorous tool for systems thinking. They are a potential tool to help you understand and visualize the complex interactions going on in your system. They are excellent at demonstrating the interconnectedness of complex systems all around us, both man-made and natural. When used in a modern tool, they provide a digital thread across a system architecture, which is exceptionally useful for identifying dependencies within our system relationships and digesting complexity into manageable chunks.

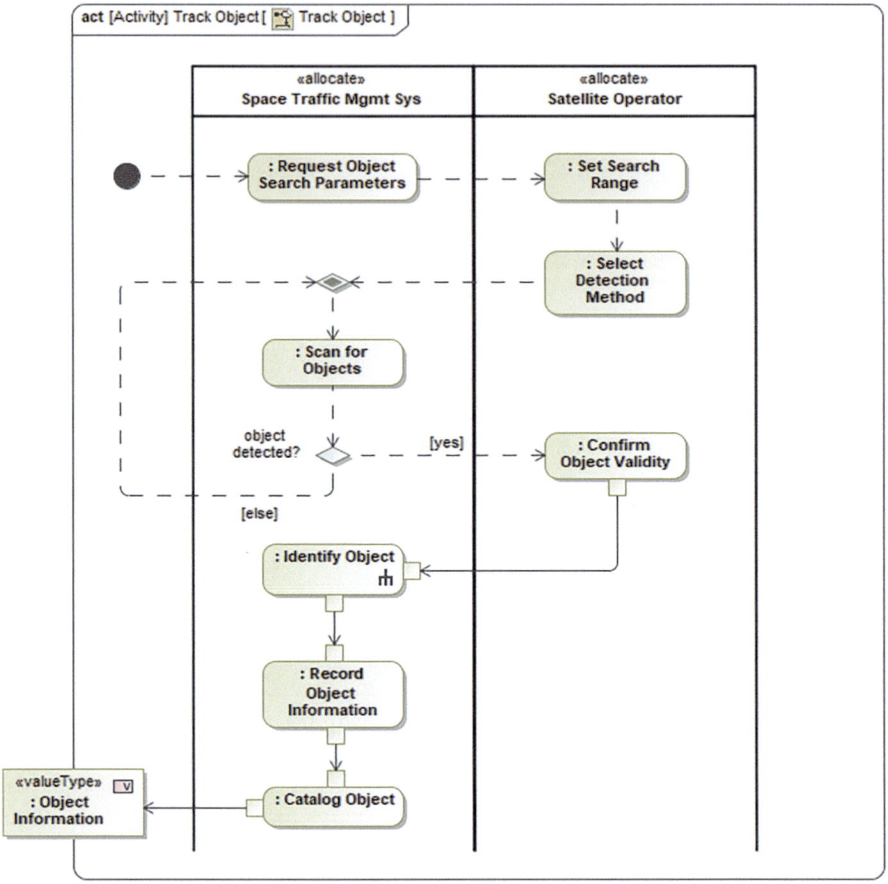

Fig. 12.7 Track object activity diagram

Note, you do not need to be in engineering or have formal training in MBSE to use these architecture models. For example, system context diagrams are a very common tool used by project teams of varying disciplines and skill levels. As long a model helps answer a question, it is useful. Here, we just presented a brief overview of a specified and well-developed way to utilize architecture models.

Should the reader wish to learn more about architecture modeling and MBSE, many excellent texts exist for teaching MBSE. *See ([5, 6, 8, 17])*. This text is not meant to teach MBSE since these texts are industry bests for learning MBSE. A great short primer on MBSE and a free copy is available [17] . Belligatti [5] and Friedenthal [8] are some of the most widely used textbooks for teaching and learning MBSE.

12.1.2 DSM as Architecture Modeling Tool

The design structure matrix (DSM) is an integral, rigorous tool for systems thinking, efficiently modeling system architectures. It helps visualize, conceptualize, and manage complex interrelationships within a system, mapping out dependencies, interactions, and feedback loops while identifying leverage points and critical elements. DSM captures a systemic perspective of technical product design, system architecture, design decisions, and analysis, particularly where subsystems have dependencies, feedback loops, and opportunities for optimization. It plays a significant role in structuring interdependencies, organizing feedback loops, inputs, and outputs in a clear, actionable way, and helping to identify areas for optimization. Model-based systems thinking (MBST) ensures that DSM inherently incorporates systems principles. By providing both tabular and graphical representations, DSM integrates systems thinking as a rigorous tool, offering rapid prototypes for effective system analysis as shown in Fig. 12.8.

We explore the space between wildfire UAV fleet deployment and UAV design, understanding their dependencies. There are multiple domains to consider such as System-Product Domain and Information Item (Specifying) Domain, Process Domain and Organization Domain as shown in Fig. 12.9 for systems analysis. The sub elements of Information Item (Specifying) Domain can be expanded to Fig 12.10. Our study includes UAV sizing for specific applications to identify leverage points (e.g., predictability, profitability, and performance). For example, a UAV size of 10 m with a wingspan of 7 m and roughly 46 kg is considered optimal. The DSM for deployment problem is defined in Fig. 12.11.

The relationship between deployment and design highlights that using smaller UAVs increases the number of launch and retrieval operations, requiring significant coordination. Thus, we need an optimal UAV size to set a proper framework. Our system dynamics model demonstrates that the right size UAV reduces deployment time. The DSM for design problem is defined in Fig. 12.12.

	Subsystem A	Subsystem B	Subsystem C	Subsystem D
Subsystem A	■		X	X
Subsystem B		■	X	X
Subsystem C			■	
Subsystem D		X		■

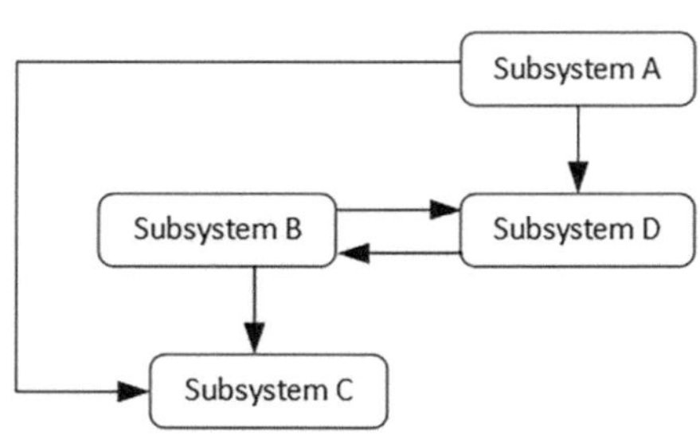

Fig. 12.8 Architectural interconnectedness using DSM

$root		1 System-Product Domain	2 UAV System	3 Air Vehicle Segment	4 Mission Payload Segment	5 Ground Segment	6 Aircraft System Ref	7 ...ormation Item (Specifying) Domain	8 Requirement Specification	9 Description	10 Drawing and Associated Lists	11 Process Domain	12 Organization Domain
⊟ System-Product Domain	1	47%											
⊟ UAV System	2		45%						10	49	25		
⊞ Air Vehicle Segment	3			29%	8	3			3	23	13		
⊞ Mission Payload Segment	4			13	8%				3	13	7		
⊞ Ground Segment	5			3		5%			1	12	3		
Aircraft System Ref	6						3%		3	1	2		
⊟ Information Item (Specifying) Domain	7							51%					
⊞ Requirement Specification	8								5%	18			
⊞ Description	9						1		5	36%			
⊞ Drawing and Associated Lists	10									3	8%		
Process Domain	11											1%	
Organization Domiain	12												1%

Fig. 12.9 DSM as system analysis

Observation for System Architecture Using DSM

In this section, we highlight the observation from the research study and connect the discussion among deployment, waypoints and UAV design trade-offs, and leverage points. Observations from this research study highlight the tremendous impacts on system architecture from a systems thinking perspective. DSM reveals that mission planning depends on UAV aerodynamic design and environmental conditions, leading to better optimization in waypoint planning and energy usage. DSM efficiently groups overlapping wildfire UAV fleet deployment objectives, reducing redundancy within the system and facilitating surveillance protocols, which saves time and resources.

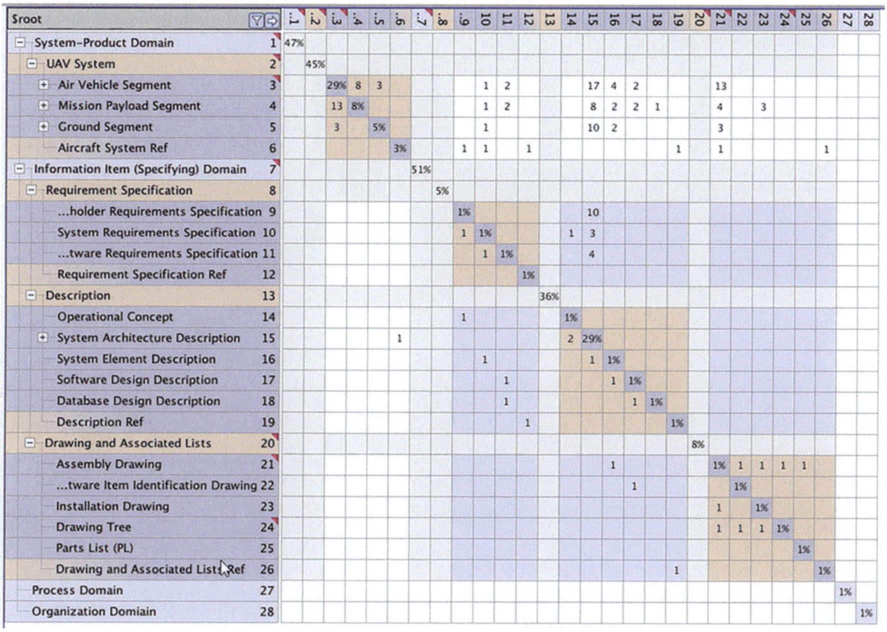

Fig. 12.10 DSM as system analysis showing sub elements

Fig. 12.11 Deployment DSM as system architecture analysis

	Mission Planning	UAV Configuration	MBS Dispatchment	Waypoints Planning	Environmental As.	UAV Launches (3x)	In-Flight Monitoring	Data Collection	UAV Retrieval (3x)	Post-Flight Analysis
Mission Planning				X						
UAV Configuration			X							
MBS Dispatchment						X				
Waypoints Planning		X								
Environmental Assessment	X			X						
UAV Launches (3x)			X		X					
In-Flight Monitoring								X		
Data Collection										X
UAV Retrieval (3x)										X
Post-Flight Analysis										

Summary for DSM as Systems Thinking Tool

The design structure matrix (DSM) is an effective way to integrate systems thinking while developing system architecture. From our research study, we conclude that

Fig. 12.12 Design DSM as systems architecture analysis

	UAV Systems Arch.	Aerodynamic Design	Structural Design	Propulsion System D.	Avionics Systems D.	Flight Controls D.	Power Systems D.	Systems Integration	Systems Validation	Documentation
UAV Systems Architecture	■	X	X	X	X	X	X			
Aerodynamic Design		■	X					X		
Structural Design			■	X			X	X		
Propulsion System Design				■			X	X		
Avionics Systems Design					■	X		X		
Flight Controls Design						■		X		
Power Systems Design							■	X		
Systems Integration								■	X	
System Validation									■	X
Documentation										■

DSM is an effective tool for implementing systems thinking in system architecture. DSM helps visualize and optimize complex interdependencies, making it easier to identify leverage points and areas for improvement. DSM as an architectural tool ensures systems thinking, and it allows systems to design operational efficiency for the case of wildfire UAV fleet deployment.

12.2 System Dynamics Models

System dynamics models are vital in systems thinking, particularly for time-sensitive systems. This approach analyzes system behavior over time, incorporating feedback loops, delays, and system responses. Introduced by pioneers like Jay Forrester, Donella Meadows, and Peter Senge, these models emphasize the importance of understanding time-sensitive behavior in complex systems. System dynamics models enable us to analyze system evolution, identify time delays and their impact on performance, and simulate the effects of interventions throughout a system's life cycle. In this chapter, we use tools like Cameo to evaluate system behavior under time and delay constraints, leveraging these principles to create models that support decision-making by predicting system responses, as Forrester's ideologies highlight.

12.2.1 MBST Extends System Dynamics to Include Other Modeling Methods

The MBST framework or process shown in Fig. 11.1 is created so that all the steps can potentially include other methods repurposed to fit systems thinking. For example, finding the reference mode behavior and identifying recognizable patterns in them, which traditionally is a very cumbersome process, need not be done manually and can be boosted by AI, as will be discussed later in this chapter. Other than system dynamic modeling that is naturally included in MBST, the other modeling methods either support the system dynamic modeling (e.g., to provide the causal relation details: equations, look up table, etc.) or work in parallel with the system dynamic modeling to enable the slow and fast learning loops of the framework.

MBST is agnostic to profession or discipline because it can be applied to any phenomenon and discipline, such as social, biological, economic, political, and technical. It can be used to solve problems ranging from predicting weather patterns, species extinction, marathon race performance, traffic flow problems, escaping the technology transition valley of death, and optimizing the operation of an innovation ecosystem.

MBST is not to be confused with model-based systems engineering (MBSE). MBSE is a framework, language, and tool for the systems engineering discipline and has recently enjoyed greater adoption in the aerospace and defense industries. Initially, its main claim to fame was architectural modeling or capturing the interconnects between the purpose, structure, and behavior of the 3-systems. MBSE is also growing to include more modeling methods, but its primary purpose, like systems engineering, is to help create complex system products and services. So, MBSE is more about developing solutions, whereas MBST is more about defining, refining, and improving our mental model of problems. As we discussed previously, understanding problems and finding solutions are two sides of the same coin, but they are not the same. As MBST and MBSE grow, some of their capabilities may overlap or use similar essential tools and theories.

12.2.2 Benefits of the MBST Versus Just ST

MBST uses models to gain understanding and behavioral prediction of complex systems in science, engineering, politics, economics, and any complex phenomena that are difficult to understand or explain. It encourages a holistic view of complex phenomena instead of focusing solely on individual components. MBST considers how elements within a system interact and influence each other. This holistic understanding helps to uncover hidden connections and dynamics. The MBST framework can be applied to science, business, academic, or social disciplines.

The increasing complexity of many of today's systems makes using model-based systems thinking necessary to explain systems operations, systems evolution, and possible systems behaviors. The explosion of the availability of mass quantities of data through open-source platforms and low-cost AI-powered platforms is an enabler to using MBST by analyzing patterns and causal relationships to understand and predict complex systems structures and behaviors. MBST is a way to translate data into actionable information.

The essence of MBST and the fundamental reason that it works are the SPs and SWs that do not require any models! So, the emphasis is on ST, and we need to make it clear that MBST does not replace ST. It requires it and enforces it!. So, we are not suggesting that model based is the only way to get the above benefits. ST is analogous to having a paper map and compass to find your way in a remote forest. But when we have a large remote forest with complex terrain and a number of teams that are distributed in the remote forest but are collaborating together, then we may wish for more advanced tools that make finding the path and team communication a lot more convenient and efficient like GPS, Live Terrain maps, and Walkie Talkies. MBST is the more convenient version of ST.

12.2.3 System Dynamics as Systems Thinking Model: Case Study of Water and Energy System in Cities

The interconnections between water and energy systems in cities are complex and multifaceted. Water is required for energy generation and energy is required for water services. Thermoelectric plants, for example, require large quantities of water for cooling, while hydroelectric plants directly use water to generate electricity. Likewise, energy is required to pump groundwater and transport treated water to homes. The more complex a city's water system is, the greater the energy demands. As populations and city systems grow, the demand for water and energy increase which in turn can increase greenhouse gas emissions if those energy sources are not carbon-free.

The interconnection and dependency of water and energy systems is an ideal systems thinking application. Applying systems thinking provides an understanding of the feedback loops and interactions between these systems. For instance, increasing energy efficiency in water systems can reduce the energy demand, allowing for more applications of renewable energy sources. Similarly, improving the efficiency of water use in energy production can help preserve water supplies for other uses, like food production.

This dynamic interaction between urban water and energy systems is a particularly pressing topic in regions like Colorado where water resources are scarce and managed closely to support both energy production and urban drinking water. If we could study these relationships, we could better understand the trade-offs between

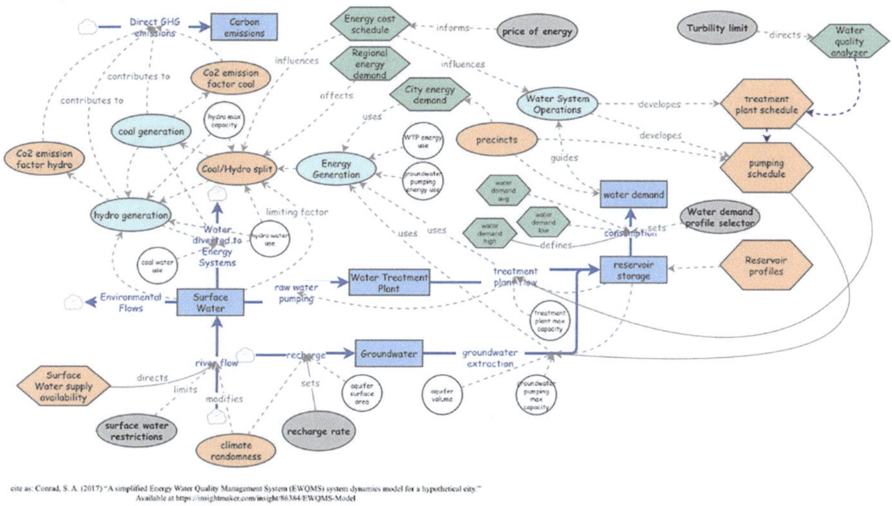

Fig. 12.13 Conrad, S. A. (2017) "A simplified Energy Water Quality Management System (EWQMS) system dynamics model for a hypothetical city." Available at https://insightmaker.com/insight/86384/EWQMS-Model

water, energy, and greenhouse gas emissions. But how specifically do these systems behave when interactions occur?

One we could use to better understand these interactions is to use causal loop diagrams to investigate water to energy causality. We could also study more specific behavior by using system dynamic modeling. One example of a system dynamics model is the EWQMS (Energy and Water Quality Management System) simulator developed by Dr. Conrad at Colorado State University shown in Fig. 12.13 and can be accessed using 12.14. The EWQMS simulator was prepared following a balanced flow approach that seeks to meet water demand in the city. Surface water from rivers and lakes is treated and groundwater is pumped to meet this demand. Energy is required to pump and treat water and water supplies are diverted to produce this energy. The main goal was to build a rigorous tool that can help explore how different systems of water and energy systems in a city are connected and how they affect each other. When creating a tool like the EWQMS simulator, it is important to consider the system's boundaries, inputs, and outputs. Inputs include things like how much water is available, how much energy is used, and the energy costs, while outputs include things like the amount of water treated and energy used. These inputs and outputs interact to create feedback loops, which determine how the system adapts and changes over time.

The EWQMS simulator is a simple but powerful tool that looks at the connections between water supply, energy, water quality, and water and energy demand. It allows users to play out different situations to how the system evolves (**evolution**

Fig. 12.14 QR code to access the EWQMS simulator

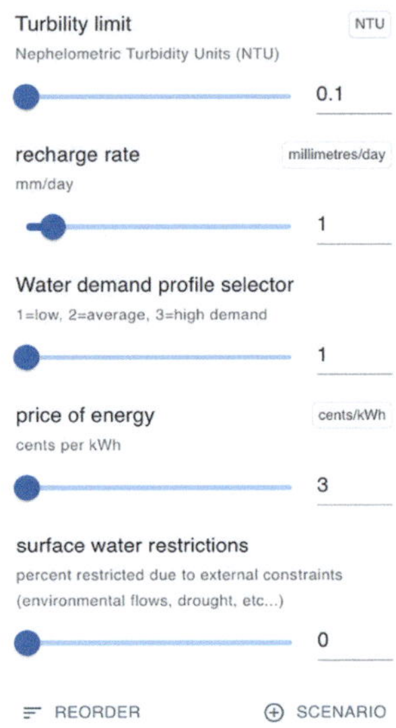

Fig. 12.15 EWQMS simulator parameter panel

principle). For example, a user can change things like energy prices, the number of particles in the water (turbidity), or how quickly groundwater recharges, and then run the simulation to see how those changes affect water treatment, groundwater levels, carbon emissions, and how the system performs overall.

The EWQMS simulator is available to use online through the insight maker platform using the QR code in Fig. 12.14. No insightmaker.com account is required to use the simulator. You can use the simulator to experiment and explore system linkages and behaviors at your own pace. You can play out a variety of scenarios such as increasing turbidity limits—a measure of how much material (like dirt, algae, or small particles) is floating in the water—or changing demand to see how the city systems respond. Below is a description of the parameters you can adjust. We encourage you to explore the system by adjusting the parameters and observing the resulting changes as depicted in Fig. 12.15. By playing around with elements like water demand profiles, energy consumption rates, and system storage capacities, you can gain a clearer understanding of the intricate balance between urban water and energy resources.

Example Case Study—EWQMS applied to water and energy in cities: The EWQMS simulator illustrates the interdependencies between urban water and energy systems using system dynamics. Here's how we can utilize the simulator to apply our systems thinking principles:

1. **Holism Principle:** The EWQMS simulator combines various components of water and energy systems. It models how water availability, energy consumption, operational costs, and greenhouse gas emissions collectively influence the overall decisions. The simulator shows that the system is more than just the sum of its parts.
2. **Architecture Principle:** The simulator shows how a change in one part of the system—like energy prices—can change how the rest of the system works. As energy prices rise, operators may change water pumping operations, which then affects energy needs and which water water supply is used.
3. **Boundaries Principle:** When designing the simulator, it was important to define what inputs and outputs to include, like water availability or energy costs. The user of the simulator can affect the boundaries depending on who's using it and what they want to learn—whether it's an operator looking to make a change or a policymaker planning for the future.
4. **Evolution Principle:** The simulator allows users to see how the system evolves when different factors change, like energy costs or groundwater recharge. This simulator shows that water and energy systems are not static and they evolve as cities grow and change.
5. **Learning Principle:** Users can gain knowledge and experience in the system by experimenting with the simulator.

By using the EWQMS simulator, we can better understand the complex interactions between water and energy in cities. It helps us make decisions that take into account the different parts of the system and how they affect each other. The EWQMS simulator supports systems thinking efforts by illustrating these relationships and making them easier to see and understand.

12.3 ABM Modeling and Stochastic Modeling as Systems Thinking Tools

Agent-based models and stochastic models are great tools that can be used for model-based systems thinking [2]. These tools can be used in a wide range of fields to study and understand systems using systems principles as a guide. An agent-based model (ABM) is a computational model used to simulate the actions and interactions of autonomous actors referred to as agents over time. Agents represent individual actors (cells, viruses, people, animals, technical components, etc.) that can come together to represent a collection of entities like organizations or groups of people or things. A unique characteristic of ABMs is that each agent operates independently based on a set of laws, rules, or behaviors. Agents interact with each other and their environment, leading to emergent behaviors that are sometimes not predictable. The primary goal of ABMs is to then discover the emerging behavior of complex systems from the rules that govern agent interactions [12].

Agent-based models are a rigorous tool because they can be used to simulate the actions and interactions of these agents to assess their effects on a system. ABMs are useful for studying complex systems where interactions between many agents can lead to phenomena that are not easy to predict. You can use ABMs with a wide range of behaviors, physics laws, and rules that make them adaptable to various fields such as science, engineering, chemistry, biology, and more.

Another strength of agent-based modeling is that it can include different types of agents, each with its unique characteristics and behaviors. This provides a system simulation that is a more realistic representation of real-world scenarios. Another plus to using ABMs as a model-based system thinking tool is that it can capture dynamic interactions over time. This can allow a modeler of a complex system to study how the system evolves and how the system responds to changes in agent behaviors and outside stimuli.

Agent-based models give the ability to perform analysis of complex systems using the scientific method of research. The scientific method is a systematic process used by scientists to explore observations, answer questions, and test hypotheses. Agent-based models give a system researcher the ability to understand the underlying element rules of a systems, predict outcomes, and test hypotheses in a controlled virtual environment, thereby enhancing the robustness and validity of the study of system dynamics. While system dynamics modeling gives a macroscopic view of a systems behavior, ABMS's provide a microscopic view of a system behavior over time. One of the strengths in using agent-based model is the ability to do counterfactual research.

Counterfactual research is used in many fields to estimate the effect of a specific action or treatment on a system. Researchers do this by comparing what actually happened to an estimate of what might have happened if the action or treatment hadn't been applied [10]. This ability to perform this type of analysis is very useful when determining the efficacy of certain elements within a system or policy effects on enterprise system operations. Common tools to generate agent-based model are NetLogo, Anylogic, etc. Figure 12.16 shows a sample interface of an agent model in Netlogo.

As mentioned above, another great tool that can be used for model-based thinking is stochastic modeling. Stochastic models are mathematical or computational models that incorporate probabilities, randomness, and uncertainty to predict several possible outcomes from an event, treatment, or behavior. Stochastic modeling is a fundamental part of statistics because it deals with probability theory, random variables, statistical inferences, and time series data analysis. A model-based thinker could use stochastic modeling to figure out the probabilities of how a complex

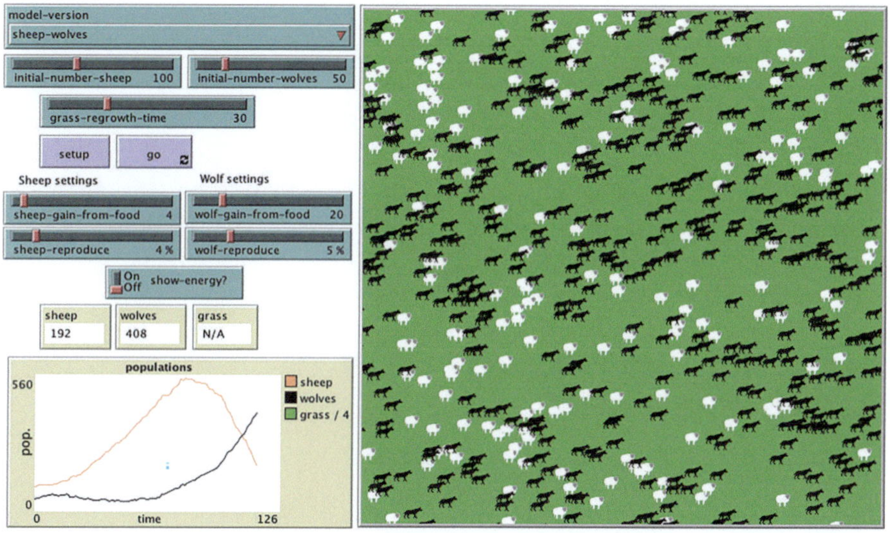

Fig. 12.16 Sample figure from NetLogo Knowledge Adoption Research

system would behave given some random interactions of its elements or figure out why a complex system behaves the way it does given some data from experiments or tests.

Some key mathematical tools that can be used for stochastic modeling are random variable analysis, Markov chains, and Bayesian analysis [24]. Systems can be defined by the behavior that emerges from their structures. These behaviors can be quantified by events. The events are associated with probabilities of occurring or not occurring or some measure between. If the event or events are unknown and have some probability of occurring, it is random. This outcome of possibilities can be assigned a variable like X or Y. If you cannot be sure of the value of X or Y, it is a random variable. Mathematical tools that can define random variables are usually some sort of distribution of possible outcomes. These include normal, binomial, Poisson, and exponential distributions, which help describe the likelihood of different outcomes.

Transitions are changes of the state of the system and the probabilities associated with the changes of state are called transition probabilities. A transition matrix for a Markov chain describes the probabilities of transitions between states. For example, in Fig. 12.18, PAB is the probability of transition from State A to State B. Each of the probabilities can be calculated using Bayesian inference as mentioned earlier. A real-world application would be the ability to predict gas prices due to changes in policies, production, refinement, and exploration.

Bayes' theorem is widely used in various fields such as statistics, machine learning, and medical diagnostics to make informed decisions based on incomplete or uncertain information. Figure 12.17 shows the theorem. It's named after Thomas Bayes, an eighteenth-century mathematician and Presbyterian minister. The

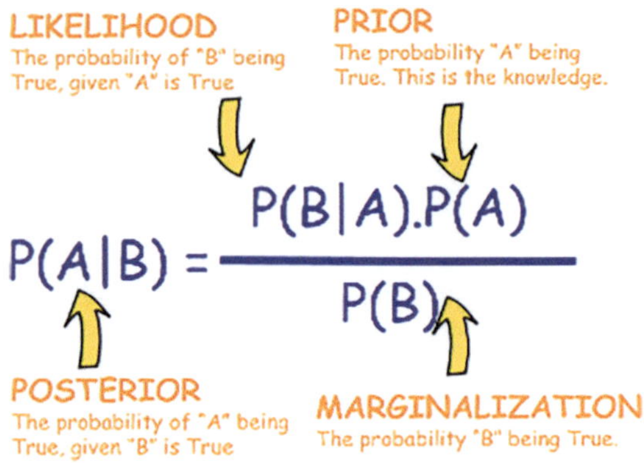

Fig. 12.17 Image of Bayes' theorem that provides a way to update the probability of a hypothesis based on new evidence

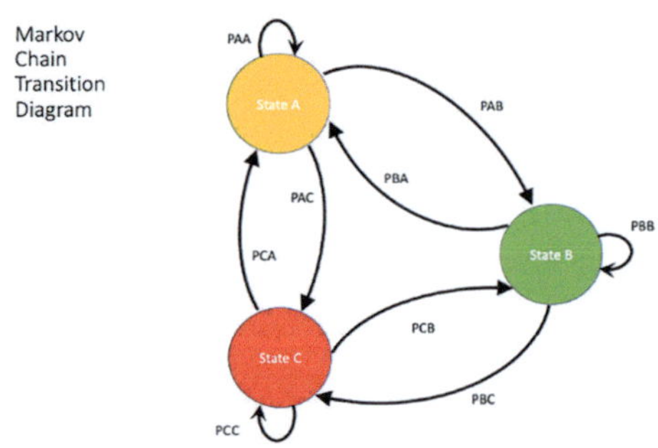

Fig. 12.18 A Markov chain or Markov process is stochastic or probabilistic random process describing a sequence of states of possible events in which the probability of each event depends only on the state attained in the previous event

theorem uses the conditional probability of an event, which is the probability that the event occurs given that another event has already occurred. Bayes' theorem is widely used in various fields such as medicine, for diagnostic testing, in finance, for risk assessment, and in machine learning, for classification tasks. It allows for a rigorous approach to the concept of "learning from data" and can be particularly powerful in situations where information is incomplete or uncertain. By continuously updating the probabilities as new data becomes available, Bayes' theorem provides a dynamic and adaptable framework for understanding the likelihood of various outcomes. It's the cornerstone of statistical inference and decision-making processes. As stated previously, Bayes' theorem allows us to update out present knowledge or belief about an event known as the "prior" with data from probabilities of what could happen known as the "likelihood" to get knew predications as an updated belief or knowledge known as the "posterior." The most powerful contribution of Bayes' theorem is that it allows us to mathematically calculate events from data when we have data from events. This is known as inverse probability in statistical circles. It is also the way Sherlock Holmes would deduce hypothesis from observations. It is elementary!

Causal inference is another statistical model that is used in model-based systems thinking. Causal inference is a mathematical way of determining "what-if" questions based on cause and effect. Correlation analysis is prevalent in almost all of today's statistical models. Correlation analysis is the dominant way that most popular machine learning algorithms operate. Model-based systems thinking teaches us that correlation is not always related to causation. Knowing that increased sales of slushy sodas match increased sales of above-ground pools in a cyclic manner throughout the year does not indicate that slushies cause pool sales to go up. If we want to find out actionable items to effect behavior changes in systems, we must know the cause of events not just correlations of events. Causal inference gives us a model-based way to do this inference. Causal inference mathematics is built on diagrams and equations of potential outcomes, effect estimates, and conditional probabilities. These diagrams and mathematical tools allow the exploration of counterfactual outcome to answer the "what if I tried this" questions that systems engineers need to understand and design specific behaviors for complex systems.

12.4 Artificial Intelligence/Machine Learning Models

Artificial intelligence/machine learning(AI/ML) algorithms, framework, and models can be very effective rigorous tools for model-based systems thinking. AI/ML simulates human intelligence via computational hardware and software and is useful in understanding and modeling complex, nonlinear systems which have multiple parameters that interact with each other over time. Currently, there are a myriad of AI algorithms and frameworks that exist that are typically applied together to solve problems. This section does not provide a comprehensive list of AI/ML tools but instead focuses on how AI/ML overall can be used as a rigorous tool in MBST.

The birth of AI came during WW2 in trying to solve a very complex systemic problem of breaking the ENIGMA code used by German U boats to communicate which enabled them to successfully sink Allied ships. The code was considered unbreakable by methods used at that time. To solve this problem, the precursor of AI was born when Alan Turing developed a computational system to solve the ENIGMA code problem and proposed the question "Can machines think?". This could be considered as one of the early applications of systems thinking to solve a complex problem. Alan Turing also developed the Turing test for computation systems, in which, if a computer could not be identified from a set of questions as not being a human, it was considered to have artificial intelligence [28].

Although researchers had developed a mathematical model of neurons during WW2, it wasn't until 1960 that a computational model known as the multilayer perceptron was developed which is the genesis of neural networks. A multilayer perceptron is a mathematical function that represents biological neurons interconnected in a layered structure as computer code. The key to the model's mathematical function is the change of weights between neurons which mimics learning [19, 22]. Neural network algorithms identify relationships in a dataset by adjusting the number and weights of interconnected nodes to create a representation that can be used to make predictions as shown in Fig. 12.19. Deep learning is a type of machine learning that uses multiple layers of interconnected nodes to

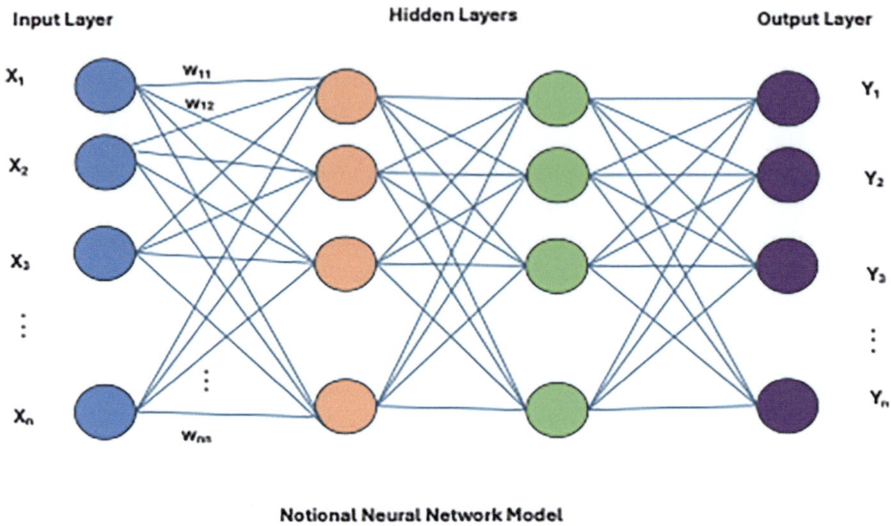

Fig. 12.19 Notional neural network model

represent complex datasets [16]. This is the foundation of current machine learning algorithms and models including graph neural networks, causal neural networks, physics-informed neural networks, and generative AI models such as generative adversarial networks(GAN), variational autoencoders (VAE), and large language models (LLM) like Chat GPT, CoPilot, Gemini, and LlAma.

AI/ML has many applications and could be useful for both model-based systems thinking, and systems engineering activities. With the growth in computational power, AI/ML has the ability to handle and find correlations in large datasets that were hidden before. These benefits can provide power and convenience, along with efficiency in data processing. However, using AI/ML also risks reducing direct human interaction and engagement with the system (**human principle**). In using AI/ML blindly, we may lose the capability to understand the causal relationships and feedback loops that underpin true systems thinking. At this point in time, AI/ML can't solely be relied on since it will precisely provide correlations but not causations. AI/ML is only as good as the data it is trained on. Garbage in is garbage out when it comes to machine learning. This is evident from biases seen in results from ML algorithms as well as unwanted behaviors of LLMs like Chat GPT. Domain knowledge is still necessary to understand correlations produced by AI/ML.

Applying AI could also reverse systems thinking progress as AI systems are good at presenting black-box solutions. AI/ML techniques take large datasets and reduce these datasets down to a workable representation of a system without specifically understanding the structure and causal relationships of the system. However, with sufficient domain knowledge and awareness, AI/ML can be used as a rigorous tool for model-based systems thinking. Consider AI/ML not as a means for simplifying the problem but as a tool for clarifying system performance.

While AI is a powerful tool, we should consider carefully how we might apply it and pair it with hands-on exploration to build robust systems knowledge. As systems thinkers, we should strive to identify the "why" behind the "what," something that AI does not provide directly. At the time of writing this text, AI is best used as an aid rather than as a replacement for systems thinking. There are several ways AI can help us gain a deeper understanding of complex systems through the efficiency of machine learning and the depth of AI analysis.

12.4.1 Systems Thinking with AI/ML

The obvious benefit of using AI/ML for ST is in automation of tasks that process large amounts of data. AI/ML could be used to apply systems principles such as the identification of the enabling systems and environments that affect a system of interest, or capture human influences on systems, or optimize key aspects of a system design where there are large datasets [15, 26].

Once we examine the use of AI for ST through the lens of systems principles, we uncover a possible disadvantage of using AI for systems thinking. It could cause the user to depend completely on AI to understand the system elements, environment, and connections which would lead to a lack of systems thinking. This dependency could prevent us from interacting directly with systems to better understand complex systems. Instead of passively relying on AI outputs, we can use AI to explore different system behaviors so we can develop a deeper understanding of system dynamics. Although machine learning algorithms are great at learning patterns of behavior and identifying system parameters from data, they cannot provide insight into system structure, causal connections, and feedback loops that are the backbone of systems thinking and system dynamics. The "black-box" nature of machine learning has been a big obstacle for its adoption in safety critical systems. There has been some work on algorithms and frameworks that enhance interpretability and understanding of how ML algorithms make decisions, but they are still limited.

Without an understanding of the structure of a system, it becomes impossible to determine the best leverage points, such as a change in mental models, to solve complex problems related to the systems. These limitations mean that even as AI improves, AI cannot replace the insight we gain (**learning principle**) from exploration of systems. We can however consider using AI tools like ChatGPT to explore scenarios to identify methods for understanding how change influences the whole system. The ability of machine learning algorithms to make future predictions is predicated on not only the amount but also the diversity of the data used to train them. Very often due to limited data and hyperparameter training, machine learning models are created that only work for a specific dataset and do not generalize well. This is both an advantage and a disadvantage of ML since it is designed to learn patterns. The problem gets worse for a complex dynamical system that changes constantly. Future predictions by ML algorithms made on such datasets will be inaccurate and useless since they rely on past behaviors which are no longer relevant.

For complex systems that require human-machine teaming, generative AI can be used to assist in ST by providing domain knowledge [4].

Currently, the best use of AI/ML in ST is to use it as a supplemental tool to enhance the results in the different steps of the MBST framework described in Chapter 11.1.

12.4.2 AI and System Dynamics

AI and SD have a fundamentally different focus since AI is about correlation and SD is about causation. A system dynamics model like AI/ML is created from understanding the patterns of behavior in data. However, unlike AI, system dynamics models utilize additional information such as studies, cumulative knowledge, and information about the environment and enabling systems. Similarly, both AI/ML and SD aim to understand how different variables affect the outcome to improve the results and make predictions of the outcome. However, machine learning accomplishes the goal, by modeling underlying patterns in the data, whereas system dynamics uses patterns in a data as well as human and domain knowledge to provide a holistic answer.

This does not mean that we cannot use AI for SD. It means that AI cannot be the only tool for SD but can be a very effective supplemental tool in SD. AI can be used to capture underlying dynamics of a system when trying to match a reference mode of a system which can be a difficult and time-consuming task for system dynamicists. LLMs can be used to capture domain knowledge about enabling systems and inform the development of a SD model [3, 9, 30].

12.4.3 AI for Architectural Modeling

AI/ML has tremendous potential for application in MBSE. It can increase the overall efficiency of the MBSE process by analyzing data generated during system development to provide insights to optimize the system design and performance. AI techniques such as ML and NLP can help automate requirements extraction, management, and validation by extracting requirements from textual sources such as contracts, regulations, and standards, and then automatically mapping them to system models. Other applications include using large language models to extract and identify human stories to help assist in development of complex systems and for risk assessment. Generative AI can be used for stakeholder analysis to understands the complex interconnections of relationships, roles, and influences of different stakeholders to inform strategies of engagement with them. Other possible uses include system design, quality management, knowledge-based validation of system design, generation of test scenarios, identification of design updates, automation of fault analysis, etc.[7, 21, 23].

Traditionally, the generation of DSMs is accomplished through manual methods which involve interviewing experts to elicit critical system elements and the relationships between them. These manual methods can be time-consuming and costly. The integration of AI techniques such as regression analysis with the design structure matrix technique can be effective in enhancing the execution and optimization of complex business processes and can save lots of productive turnaround time (TAT) which equates to reduced costs. LLMs can be used to support the generation of DSM by generating workflows and improve productivity. LLMs can also be used for automatic generation of design structure matrices as the product and processes evolve over time to provide significant cost savings [11, 14, 20].

12.4.4 AI for Agent Modeling and Bayes/MCMC

Earlier we introduced ABM which is a methodology used to build formal models of real-world systems that are made up by individual units (agents) which repeatedly interact among themselves and/or with their environment. Each agent has its own state and rules that govern how they behave and update their state to interact with other agents or the environment.

Machine learning can be used to derive agent behaviors for development of ABMS and are referred to as ML-driven ABMs. AI/ML can be used in eliciting the behavioral rules, defining learning and adaptability in agents and testing the sensitivity and validity of the model outputs. Machine learning can be used to automatically build agent-based models from empirical data. This is useful in designing policies and practices for behavior change across multiple domains since linear statistical techniques do not capture the interconnected relationships between the determinants of the behaviors and their impacts on their outcomes [13, 25, 27].

Bayes inference is a critical building block for AI. It enables AI systems to model and reason about uncertainty by updating beliefs based on new evidence. A Bayesian network is a type of graphical model that uses probability to determine the occurrence of events and derive relationships between multiple variables or events. These relationships are derived based on joint and conditional probabilities [18].

In the previous section, we introduced the concept of Markov chains. AI Markov chain Monte Carlo (MCMC) algorithms leverage Markov chains to obtain numerical approximations which are vital in addressing problems related to Bayesian inference, model fitting, and other statistical analyses in AI applications. Markov chain Monte Carlo (MCMC) is a computational iterative technique used to obtain a sequence of random samples from a complex probability distribution. It plays a critical role in AI and in complex modeling and inference tasks due to its ability to handle dimensionality and complex probability distributions and quantify and incorporate uncertainty in AI models. It can be computationally intensive for large datasets and high-dimensional models and may encounter convergence issues in certain scenarios which can make it challenging to use in real-time decision-making and resource-constrained environments. Real-world applications range from cancer, environment, geophysics, and robotics to cryptography [1, 29].

12.5 Reflection on Chapter 12 Material

In this chapter, we focused on the rigorous tools used in the model-based systems thinking (MBST) method. MBST relies heavily on various modeling methods that describe the purpose, structure, behavior, structure, and purpose of complex systems and their interrelatedness. These models serve as crucial tools for understanding the interactions within and between the 3-systems. We showed various real-world applications of these modeling methods, without going into too much detail, to plant a seed in the reader's minds in case they want to add them to their skill sets. We have touched on the following modeling methods in this book: system dynamics, architectural modeling, design structure matrix (DSM), agent-based modeling, stochastic modeling, AI/machine learning modeling, and optimization.

These are currently among the most helpful modeling methods under the MBST umbrella method, but this list is not intended to be exhaustive. These modeling methods classically require time and effort to learn and apply. System dynamics modeling is a powerful tool but takes a relatively long time to learn how to apply it effectively. However, this situation is temporary. LLMs and AI tools will soon boost all the major steps of the slow and fast learning loops of MBST. For example, our recent research activity used LLMs to create excellent quality causal loop diagrams that were at least as good or better than 60 graduate students taking a systems thinking class. We have seen enough research progress to believe that creating modeling artifacts from scratch, gathering relevant reference mode data, fine-tuning the models to match the reference mode data, and all the other major steps of the MBST method in Fig. 11.1 will eventually be significantly boosted by AI tools. So, we expect the current high barrier to entry into these various modeling methods will disappear soon.

References

1. Aghaei, T. V., et al. (2021). A real-world application of Markov Chain monte carlo method for bayesian trajectory control of a robotic manipulator. *ISA Transactions, 125*, 580–590.
2. Antelmi, A., et al. (2023). Experimenting with agent-based model simulation tools. *Applied Sciences (Switzerland), 13*(1). https://doi.org/10.3390/app13010013
3. Armenia, S., et al. (2024). Zooming in and out the landscape: Artificial intelligence and system dynamics in business and management. *Technological Forecasting and Social Change, 200*, 123131.

4. Barcaui, A., & Monat, A. (2023). Who is better in project planning? Generative artificial intelligence or project managers? *Project Leadership and Society, 4*, 100101.
5. Belligatti, L. (2013). *SysML distilled : A brief guide to the systems modeling language*. Pearson Education.
6. Borky, J., & Bradley, T. (2019). *Effective model-based systems engineering*. Springer.
7. Chami, M., et al. (2022). Artificial intelligence capabilities for effective model-based systems engineering: A vision paper. In *Special Issue: 32nd Annual INCOSE International Symposium* (Vol. 32, No. 1, pp. 1160–1174).
8. Friedenthal, S., Moore, A., & Steiner, R. (2012). *A practical guide to SysML: The systems modeling language* (2nd ed.). Morgan Kaufmann. ISBN: 978-0-123852-06-9.
9. Gadewadikar, J., & Marshall, J. (2024). A methodology for parameter estimation in system dynamics models using artificial intelligence. *Systems Engineering, 27*, 253–266.
10. Gerstenberg, T., et al. (2021) A counterfactual simulation model of causal judgments for physical events. *Psychological Review, 128*(5). https://doi.org/10.1037/rev0000281
11. Gopsill, J. A., et al. (2016). Automatic generation of design structure matrices through the evolution of product models. *Artificial Intelligence for Engineering Design, Analysis and Manufacturing, 30*(4), 424–445.
12. Grimm, V., et al. (2006). A standard protocol for describing individual-based and agent-based models. *Ecological Modelling, 198*(1–2), 115–126. https://doi.org/10.1016/j.ecolmodel.2006.04.023
13. Ionescu, Ş., et al. (2024). Exploring the use of artificial intelligence in agent-based modeling applications: A bibliometric study. *Algorithms, 17*, 21.
14. Koh, E. C. Y. (2024). Auto-DSM: Using a large language model to generate a design structure matrix. *Natural Language Processing Journal, 9*, 100103.
15. Kuo, Y. H., et al. (2020). An integrated approach of machine learning and systems thinking for waiting time prediction in an emergency department. *International Journal of Medical Informatics, 139*, 104143.
16. LeCun, Y., Bengio, Y., & Hinton, G. (2015). Deep learning. *Nature, 521*, 436–444.
17. Long, D. A., & Scott, Z. (2011). *A primer for model-based systems engineering* (2nd ed.). Vitech Co.. ISBN: 978-1-105588-10-5.
18. Mark, C., Metzner, C., Lautscham, L., et al. (2018). Bayesian model selection for complex dynamic systems. *Nature Communications, 9*, 1803.
19. McCulloch, W. S., & Pitts, W. (1943). A logical calculus of the ideas immanent in nervous activity. *Bulletin of Mathematical Biophysics, 5*, 115–133.
20. Medoh, C. N., & Telukdarie, A. (2018). A methodology to integrate artificial intelligence with the design structure matrix approach. In: *IEEE International Conference on Industrial Engineering and Engineering Management (IEEM)*, Bangkok (pp. 783–787).
21. Mordoch, A., & Gabel, A. (2023). Future consideration of generative artificial intelligence for systems engineering design. In *NDIA Conference Proceedings*.
22. Rosenblatt, F. (1957). *The Perceptron, a Perceiving and Recognizing Automaton Project Para*. Technical Report. Cornell Aeronautical Laboratory (pp 460–461).
23. Savary-Leblanc, M. (2019). Improving MBSE tools UX with AI-empowered software assistants. In *ACM/IEEE 22nd International Conference on Model Driven Engineering Languages and Systems Companion (MODELS-C)*, Munich (pp. 648–652).
24. Shone, R., Glazebrook, K., & Zografos, K. G. (2021). Applications of stochastic modeling in air traffic management: Methods, challenges and opportunities for solving air traffic problems under uncertainty. *European Journal of Operational Research, 292*(1), 1–18. https://doi.org/10.1016/j.ejor.2020.10.039
25. Taghikhah, F. R., et al. (2022). Machine-assisted agent-based modeling: Opening the black box. *Journal of Computational Science, 64*, 101854.
26. Tang, Y., et al. (2020). Focus issue: When Machine learning meets complex systems: Networks, chaos, and nonlinear dynamics. *Chaos, 30*, 063151.

27. Turgut, Y., & Bozdag, C. E. A. (2023). A framework proposal for machine learning-driven agent-based models through a case study analysis. *Simulation Modelling Practice and Theory, 123*, 102707.
28. Turing, A. M. (1950). Computing machinery and intelligence. *Mind, 59*, 433–460.
29. Wang, F., Idrees, M., & Sohail, A. (2022). AI-MCMC for the parametric analysis of the hormonal therapy of cancer. *Chaos, Solitons & Fractals, 154*, 111618.
30. Yin, R. (2020). *Applying Machine Learning to Learn System Dynamics Models for Urban Systems*. MA Thesis. TU Delft.

Systems Thinking Glossary for Beginners

System A *System* is a *Set* of Interconnected elements that together form a *Structure* and exhibit *Behavior* that fulfill a *Purpose* as a Whole.

System of Interest (SoI) *System-of-Interest (SoI)* is the system that we are responsible for understanding, creating, developing, improving, buying, selling, or maintaining.

Enabling Systems Enabling Systems are the systems (outside of the *SoI*) that support the design, development, operation, maintenance, and other phases of the life cycle of the *SoI*.

Context System or System Context System Context is the environment or the system that the S*ystem of Interest (SoI)* fits into.

3-Systems Systems thinking encourages us to balance our efforts and focus among a minimum of 3-Systems (*SOI*, *Enabling System*, and *Context System*) to increase our chances of understanding complex real-world issues and finding the best leverage on them.

System Architecture The *Purpose, Structure,* and *Behavior* of a system influence each other and define its *Architecture*.

Systems Boundaries *Systems Boundaries* separate all the systems relevant to understanding and dealing with a problem. For example, the boundaries of the *3-Systems* show how they are separate yet interconnect or interact with each other.

System of Systems *System of Systems (SoS)* is a group of systems that are significantly independent of each other (in terms of ownership, management, location, etc.), but they collaborate, compete, or integrate to achieve a greater common purpose. For example, US Defense is established through the US Space Force Systems, US Air Force Systems, US Army Systems, US Navy Systems, and many other systems that have clear separation boundaries and operate separately but collaborate to achieve an overall defense capability.

K. Eftekhari Shahroudi et al., *Practical Systems Thinking*,
https://doi.org/10.1007/978-3-031-89394-0

Patterns of Behavior *Patterns of Behavior* are the regular, recognizable, or repeatable ways a system appears, operates, moves, or changes over time, such as reinforcing, balancing, goal-seeking, oscillating, overshooting, and collapsing behaviors etcetera

Emergent Behavior or **Emergence** *Emergence* refers to the phenomenon where the overall system exhibits patterns of behavior and properties that are not apparent, predictable, or simply discoverable from the characteristics of the individual elements alone.

Complexity *Complexity* in a system arises from the number and nature of interconnections, interdependencies, and relationships among the *elements* of the system and among the system with other systems relevant to the system.

Complex System A *Complex System* is a system whose emergent behavior is difficult to predict. It is characterized by an interconnected set of parts that interact, leading to emergent behaviors that are not simply or linearly predictable from the characteristics of the parts.

Systems Principles (SPs) *Systems Principles* are reliable and practical systems-oriented principles that are universally true. They guide our thinking, action, and methods to help us view, understand, and deal with real-world problems as systems. The SPs and their application are covered in detail in Part II of this book, and listed briefly in the next section.

Systems Thinking (ST) *Systems Thinking* is an inter-relational trans-disciplinary *Philosophy* and *Method* for *Understanding* real-world (Complex) problems and improving the odds of success of the *Systems* that may *influence* them.

The "S" in Systems Thinking (ST) Reason for the plural form is that ST always considers ALL the systems that are relevant to a problem and how they interact. The focus is never inside just one system in isolation as the 3-Systems Principle implies.

Linear Compartmentalized Thinking *Linear Compartmentalized Thinking* is the classical approach to understanding and dealing with a problem by dividing it into smaller parts and focusing our effort inside them. It is based on the expectation that the cause of problems is near (in time and space) to where the problem is observed, and that if we understand and improve the internals of parts, then the whole will naturally improve when the parts are put together.

Ways of the Systems Thinker (or **Systems Ways (SWs)**) The *Ways of the Systems Thinker* are suggestions to help us align our thinking, approach, decisions, and behaviors with the Systems Principles.

Rigorous Method A method is rigorous if it (by design, framework, process, or machinery) reliably and repeatably enforces the disciplined application of principles, laws, or knowledge for some identified purpose.

Rigorous Systems Thinking Methods Rigorous ST Methods reliably enforce the application of SPs to real-world problems. Examples are Causal Loop Diagrams, System Dynamics, Machine Learning, Agent-Based Modeling, Stochastic Modeling, Design Structure Matrix (DSM), etc., that can be set up explicitly to enforce the application of SPs. These are introduced in Part II and covered in more detail in Part III of this book.

Model-Based Systems Thinking Model-Based Systems Thinking (MBST) is a method (including framework, languages, and models) that integrates various *rigorous ST methods* to augment our ability to view, understand, and deal with real-world problems as systems.

Systemic Leverage *Systemic Leverage* attempts to identify the most useful and practical locations within the architectures of the *3-Systems* to focus our efforts for better outcomes.

Systemic Vs Systematic Methods The former is more about applying SPs and SWs to *Understand* Real World Problems by transforming them to systems problems. The latter is more about *Developing Solutions* (Framework, Tools, Process, or Machinery) to deal with systems problems. Like two sides of the same coin, they are very different, but they coexist and need each other to be successful.

Mental Models *Mental Models* are perceptions and mental representations of real-world problems that impact human actions, behaviors, and solutions.

Systems Thinking Culture An organization or team culture that fosters systems thinking exhibiting common mental models characterized by applying SPs, SWs, and Rigorous ST methods to understanding, dealing with, and creating solutions for real-world problems.

System Life Cycle *System Life Cycle* refers to the series of stages that a system goes through from its inception to its retirement or decommissioning, analogous to the various life cycle stages of a frog.

Management Flight Simulator (MFS) A *management flight simulator* is a game-like simulation that enables decision makers to understand and see the impact of their decisions and policies on the patterns of behavior of complex socio-technical systems.

Systems Principles (SPs)

Holism Principle: (Complex) *Systems* are (much) *more* than the *sum* of their *parts—elements*, *interconnects*, and *purpose*.

Architecture Principle: The *purpose*, *structure*, and *behavior* of systems *influence* each other.

3-Systems Principle: *Systems* come in (at least) *three—context systems*, *systems of interest (SoI)*, and *enabling systems*.

Boundaries Principle: *Systems boundaries* depend on *perspectives*, *purposes*, and *roles*.

Evolution Principle: Systems have *life cycles* and they *evolve*.

Emergence Principle: *Interactions*(including feedback) within and between the three systems cause (sometimes unwanted) *emergent behaviors*.

Learning Principle: (Common) *Mental models* influencing human *decisions* about *architectures* are *learnable* and *adaptable*.

Human Principle: *Humans* impact and are impacted by the (wanted/unwanted) *emergence* of *all* reachable systems.

K. Eftekhari Shahroudi et al., *Practical Systems Thinking*,
https://doi.org/10.1007/978-3-031-89394-0

Index

K. Eftekhari Shahroudi et al., *Practical Systems Thinking*,
https://doi.org/10.1007/978-3-031-89394-0

R

S

MIX
Papier aus verantwortungsvollen Quellen
Paper from responsible sources
FSC® C105338

If you have any concerns about our products,
you can contact us on
ProductSafety@springernature.com

In case Publisher is established outside the EU,
the EU authorized representative is:
Springer Nature Customer Service Center GmbH
Europaplatz 3, 69115 Heidelberg, Germany

Printed by Libri Plureos GmbH
in Hamburg, Germany